Lecture Notes in Chemistry

Edited by G. Berthier, M. J. S. Dewar, H. Fischer
K. Fukui, H. Hartmann, H. H. Jaffé, J. Jortner
W. Kutzelnigg, K. Ruedenberg, E. Scrocco, W. Zeil

9

A. Julg

Crystals as Giant Molecules

Springer-Verlag
Berlin Heidelberg New York 1978

Author

André Julg
Université de Provence
Laboratoire de Chimie Théorique
Place Victor Hugo
F-13331 Marseille Cedex 3

Library of Congress Cataloging in Publication Data

Julg, André, 1926–
    Crystals as giant molecules.

    (Lecture notes in chemistry ; v. 9)
    Bibliography: p.
    Includes indexes.
    1. Crystals. 2. Molecules. 3. Molecular crystals.
I. Title.
QD921.J8   548   78-10489

ISBN-13: 978-3-540-08946-9   e-ISBN-13: 978-3-642-93099-7
DOI: 10.1007/ 978-3-642-93099-7

2152/3140-543210

Introduction

Traditionally, when one deals with crystals, the first property to be presented is the periodicity of the lattice, and all methods of study are based on this characteristic, which is considered essential. In fact, crystals differ from the molecules of finite size that are studied in chemistry, only in their extremely large number of particles. Furthermore, the existence of faces, which limit the spread of crystals in space, necessarily breaks the periodicity of the system. For these reasons it is natural to apply to crystals the concepts and methods that have been widely tested in the study of molecules. Pauling first emphasized this point [1] and used it to explain the electronic structure of crystals, thought to be infinite and perfect.

The aim of this work is to show, with the help of a few examples, the possibilities offered by quantum chemistry for tackling the problems of crystal electronic structure, of crystallographic arrangements as well as their macroscopic shape, and of distortion effects caused by the presence of faces. The area related to the existence of energy bands (allowed or forbidden), gap, electric, magnetic or optical properties will not be touched upon.

Because the present work is based on the structural analogies that exist between crystals and molecules, some general results for molecules will first be briefly reviewed. After showing how these notions can be transposed to crystals, a scheme of structural classification of crystals will be presented. This will permit a subsequent logical treatment of the problem of crystallographic arrangements in infinite networks.

In addition to results already published (subsequently quoted in references), the present work also contains a few original ideas and unpublished results.

To facilitate the reading, some information and mathematical developments have been transfered to the Appendix.

Acknowledgements

The author is indebted to many people who made this text possible. He especially wishes to thank Prof. G. Berthier and Prof. E. Scrocco who have given him an occasion to write this book, Mrs. M.F. Fiori (Centre National de la Recherche Scientifique) for typing the manuscript and the final version, Mrs. P. Fried for the language corrections, my colleague G. Öttl for the original drawings of the crystal lattices.

Finally, I thank Mrs. O. Julg, my wife, for help in reading and correcting the final version.

Contents

I. General Results Concerning Molecules

1. The Molecular Orbitals

If nuclei are assumed to be infinitely heavier than electrons, the problem of molecular structure is reduced to determination of electron behavior in the electrostatic field of the nuclei, which is thought to be fixed. This is the Born-Oppenheimer approximation [2], which will be used here throughout.

Consider the total multielectron wave function of the system: $\Psi(1,2,...n)$ and its energy E ($1,2,...n$ represents the set of coordinates x_i, y_i, z_i for the electrons $\underline{i}$). The Schrödinger equation, corresponding to stationary states of the molecule,

$$H\psi = E\psi \qquad (1)$$

where H is the Hamiltonian operator of the system, cannot be solved directly. Thus, as approximate solutions, one looks for determinants built up on space functions φ_i — corresponding to a <u>molecular orbital</u> — which depends only on the coordinates of one electron, multiplied by one of the spin functions — α or β — which correspond to the eigenvalues 1/2 and -1/2 of the S_z operator, respectively. Indeed, such a structure for ψ is the simplest that conserves the equivalence between the electrons and ensures a change of sign for the wave function when two arbitrary electrons are exchanged. For the ground state of a molecule with an even number of electrons, one determinant will be sufficient in almost all cases:

$$\psi(1,2,...n) = \frac{1}{\sqrt{n!}} \begin{vmatrix} \varphi_1(1)\,\alpha(1) & \varphi_1(1)\,\beta(1) & \varphi_2(1)\,\alpha(1) & \varphi_2(1)\,\beta(1) & \cdots \\ \varphi_1(2)\,\alpha(2) & \varphi_1(2)\,\beta(2) & \varphi_2(2)\,\alpha(2) & \varphi_2(2)\,\beta(2) & \cdots \\ \cdot\cdot\cdot\cdot\cdot\cdot\cdot\cdot\cdot\cdot\cdot\cdot\cdot\cdot\cdot\cdot\cdot\cdot\cdot\cdot\cdot\cdot\cdot \\ \varphi_1(n)\,\alpha(n) & \varphi_1(n)\,\beta(n) & \varphi_2(n)\,\alpha(n) & \varphi_2(n)\,\beta(n) & \cdots \end{vmatrix} \qquad (2)$$

Each molecular orbital φ_i is used twice, successively associated with each of the spin functions α and β.

To determine the functions φ_i, one writes that the energy related to the function ψ:

$$\overline{E} = <\psi H\psi> \qquad (3)$$

is minimal. To obtain φ_i explicitly, it is convenient in pratice to expand these functions as a set of functions ξ:

$$\varphi_i = \sum_r c_{ir} \, \xi_r \qquad (4)$$

In this development, the set ξ must be theoretically complete. However, many calculations on molecules have shown that a good description of their electronic structure is obtained using, for the functions ξ, the <u>atomic</u> <u>orbitals</u> (χ) that are used in the most elementary description of both the ground state and the first excited states of isolated atoms:

$$\varphi_i = \sum_r c_{ir} \, \chi_r \qquad (5)$$

e.g., in the case of the hydrogen atom, the orbital <u>1s</u> ; for a carbon or an oxygen atom, the orbitals <u>1s</u>, <u>2s</u>, and the <u>2p</u>'s ; for a titanium or an iron atom: <u>1s</u>, <u>2s</u>, the <u>2p</u>'s, <u>3s</u>, the <u>3p</u>'s, <u>4s</u>, the <u>3d</u>'s and the <u>4p</u>'s. This is the so-called <u>minimal</u> <u>basis</u>.

After determining the basic functions, the coefficients c_{ir}, introduced in relation (5), must be found. The relations for the minimization of the ground state energy with respect to these coefficients are:

$$\left\{ \frac{\partial E}{\partial c_{ir}} = 0 \quad (r) \right. \qquad (6)$$

Solving these equations and taking into account the normalization condition of the φ_is:

$$< \varphi_i | \varphi_i > = 1 \qquad (7)$$

we obtain the unknown coefficients. The expression of the energy as a function of the molecular orbitals φ_i or of the coefficients c_{ir} is given in Appendix A.

The coefficients c_{ir} are determined in practice by the Roothaan <u>self-consistent</u> <u>field</u> method [3]. A set of space functions φ_i belonging to the irreducible representations of the molecular symmetry group is obtained.

Partial densities correspond to the orbitals φ_i:

$$\rho_i = | \varphi_i |^2 \qquad (8)$$

These densities are distributed throughout the molecule, since in general the atomic functions χ_r of all the atoms are involved in their expression. The total electron density is written :

$$\rho = 2 \sum_i \rho_i \qquad (9)$$

In the determinant (2), associated with the ground state, each function φ_i appears twice, multiplied either by the spin function α or β. Therefore any unitary transformation applied to the functions φ gives a new set φ', different from the set φ, but the total wave function ψ remains unchanged [4]. The quantic description of the system is not changed. Therefore, the functions φ_i, mathematical artifices introduced in order to build up the total wave function, are completely undetermined, and there is no reason to give them any physical meaning [5]. The same applies to the corresponding densities ρ_i. From a quantic point of view, the system has an infinite number of completely equivalent descriptions. This situation, at first sight paradoxic, is due to the structure that has been imposed a priori on the total wave function and probably has no physical meaning.

2. Localization of Molecular Orbitals and the Hybridization Concept

Taking advantage of the undetermined nature of the φ_i functions, one can attempt to obtain a set φ'_i such that the corresponding partial densities $|\varphi'_i|^2$ are concentrated as well as possible in domains as confined as we want. For example, we can impose the condition that

$$\sum_i < |\varphi'_i(\mu)|^2 \frac{1}{r_{\mu\nu}} |\varphi'_i(\nu)|^2 > \qquad (10)$$

(which represents the sum of the various repulsions between electron pairs ($\mu\nu$) using each orbital φ'_i) be a maximum [6]. These orbitals φ'_i, of course, no longer belong to the irreducible representations of the molecular symmetry group. In return, they will reflect the equivalences between atom couples, as does classical chemistry.

To illustrate these fundamental considerations, let us take as example the methane molecule: CH_4. The self-consistent field method gives five molecular orbitals:

$$
\left\{
\begin{aligned}
\varphi_1 &= a\,k + b\,s + c\,(h_1 + h_2 + h_3 + h_4) \\[2mm]
\varphi_2 &= a'k + b's + c'(h_1 + h_2 + h_3 + h_4) \\[2mm]
\varphi_3 &= a''p_x + c''(h_1 + h_2 - h_3 - h_4) \\[2mm]
\varphi_4 &= a''p_y + c''(h_1 - h_2 + h_3 - h_4) \\[2mm]
\varphi_5 &= a''p_z + c''(h_1 - h_2 - h_3 + h_4)
\end{aligned}
\right. \tag{11}
$$

where $\underline{k}$, $\underline{s}$, p_x, p_y, p_z are the $\underline{1s}$, $\underline{2s}$, $2p_x$, $2p_y$, $2p_z$ atomic orbitals of the carbon atom and h_i, the $\underline{1s}$ orbital of the atom H_i ($i = 1,2,3,4$).

It is clear that the various densities $|\varphi_i|^2$ are distributed around the five nuclei of the molecule. In addition, we have only three orbitals equivalent to a rotation: φ_3, φ_4 and φ_5.

By a suitable unitary transformation, we obtain new molecular orbitals that in practice reduce to:

$$
\left\{
\begin{aligned}
\varphi_1' &= k \\[2mm]
\varphi_2' &= \frac{A}{2}\,(s + p_x + p_y + p_z) + Bh_1 \\[2mm]
\varphi_3' &= \frac{A}{2}\,(s + p_x - p_y - p_z) + Bh_2 \\[2mm]
\varphi_4' &= \frac{A}{2}\,(s - p_x - p_y + p_z) + Bh_3 \\[2mm]
\varphi_5' &= \frac{A}{2}\,(s - p_x + p_y - p_z) + Bh_4
\end{aligned}
\right. \tag{12}
$$

The expression of the last four orbitals $(\varphi_2', \varphi_3', \varphi_4', \varphi_5')$ suggests replacement of the four basic atomic orbitals: $\underline{s}$, p_x, p_y and p_z, by the four orthonormalized combinations:

$$
\left\{
\begin{aligned}
t_1 &= 1/2\,(s + p_x + p_y + p_z) \\[2mm]
t_2 &= 1/2\,(s + p_x - p_y - p_z) \\[2mm]
t_3 &= 1/2\,(s - p_x - p_y + p_z) \\[2mm]
t_4 &= 1/2\,(s - p_x + p_y - p_z)
\end{aligned}
\right. \tag{13}
$$

These orbitals have the advantage of being equivalent whereas the s and p orbitals are different. Starting from any one of them, a rotation of 109.5° around the carbon nucleus will give all of the others [7]. These new orbitals (t_i) are the so-called hybrid atomic orbitals [8]. Each corresponding density t_i^2 is concentrated around a point other than the carbon nucleus and is located toward the corresponding H_i atom. Therefore they are strongly directed [7,8,9], and the densities $|\varphi_2'|^2, \ldots, |\varphi_5'|^2$ are concentrated between the carbon nucleus and one of the hydrogen nuclei ($H_1, \ldots H_4$), respectively.

Hence, owing to a change of basic atomic function, we obtained a description similar to the general concepts of classical chemistry: one 1s orbital corresponds to the inner pair of the carbon atom and four equivalent orbitals localized in parts of space where bond electron pairs are located. In fact, the situation is not so simple. In a quantic description, all electrons have the same characteristics and the same mean position in space. The localization of molecular orbitals is a purely mathematical operation that is unrelated to the localization of electrons. The bond dashes of classical chemistry (e.g., C-H) do not represent localized electron pairs but are symbols of a possible description (among an infinity of others) of the molecule, in terms of localized molecular orbitals [10].

For a molecule, the search for maximal localization always leads to molecular orbitals that can be identified as inner atomic orbitals of isolated atoms, and in general to molecular orbitals built up upon hybrid couples pointing at one another, [7,11] which are called bonding orbitals. Sometimes, we also obtain localized molecular orbitals reduced to a single atomic orbital, usually hybrid. Such orbitals, in which no bond results, are called non-bonding orbitals. They correspond to the lone pairs known from classical chemistry (e.g., NH_3, OH_2). The situation is the same in the case of some complex molecules involving transition elements, with more or less used orbitals.

Furthermore, in planar molecules certain orbitals cannot be well localized, e.g., π molecular orbitals in conjugated molecules such as benzene. The corresponding densities remain delocalized throughout the whole molecule, as represented by the scheme:

Finally, an uncommon case, but one that will be discussed again, is that of nonplanar molecules, which have partially delocalized orbitals, i.e., orbitals involving only certain atoms. An example of this is B_2H_6, in which, besides the two inner 1s orbitals of the boron atoms and the four orbitals localized in areas corresponding to the B-H bonds on the figure, there are also two molecular orbitals consisting of a hybrid of each boron atom and the 1s orbital of the hydrogen in a bridge position. Classical chemistry calls it a three-center bond, shown with the dotted lines, in the figure:

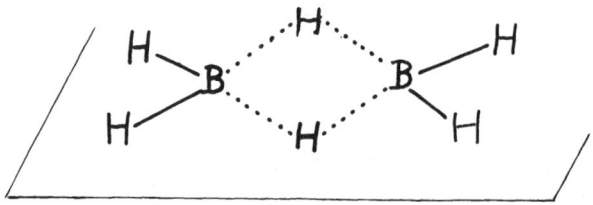

Another interesting example is the ferrocenes, in which the metallic atom lies, as in a sandwich, between two molecules having π-delocalized orbital systems. The system can no longer be described in terms of localized molecular orbitals: it must be considered as a whole.

3. Shapes of Some Usual Hybrid Atomic Orbitals

Without studying in detail the properties and shapes of various hybrid orbitals, we shall review the essential points. This subject is treated more completely in other works [7,8,9,11], to which the interested reader is referred.

If only s and p orbitals are taken into account, there are three possible types of hybridization patterns for atoms, according to the number of atomic orbitals utilized for the construction of all the hybrids:

a) sp3 hybridization:

The set of hybrids is built up from the four basic orbitals s, p_x, p_y, and p_z. There are four hybrids t_i whose corresponding centers of gravity of electron densities do not coincide with the nucleus. These hybrids point in four spacial directions (principal direction). In the methane molecule, the four directions form a regular sheaf. In a molecule with different ligands, the figure will be slightly distorted. For example, instead of 109.5° in CH_4, the angles $\widehat{HNH}$ in NH_3 are about 107° (the lone pair playing the role of a ligand). In OH_2, the angle $\widehat{HOH}$ is about 105°. The distortions are weak and can therefore be neglected in a pure qualitative description, such as is presented below.

b) sp$_2$ hybridization:

Only two p orbitals are used with the s orbital. The third p orbital remains such as it is. The three hybrid orbitals point in three coplanar directions. If the three ligands are identical, the figure formed by the directions of the orbitals is regular, with angles of 120° (BF_3). Generally, slight angular distortions of a few degrees appear, as in ethylene. They will be neglected here. The nonhybridized p orbital axis is perpendicular to the plane formed by the principal directions of the three hybrids.

c) sp hybridization:

Only one of the p orbitals is used with the s orbital to build up the hybrids. The two resulting orbitals point in two opposite directions. The axes of the two remaining p orbitals are perpendicular to each other and to the axis common to the hybrids.

d) Hybridization using d orbitals:

If, in addition to s and p orbitals, d orbitals are also taken into account, the following fundamental combinations result:

α) sp_3d_2: six hybrids point to half-axes of a trirectangular coordinate system. These hybrids have equivalent cubic symmetries. If two such hybrids, pointing in opposite directions, are equivalent to each other but are different from the other four, quadratic (also called tetragonal) symmetry is the result.

β) sp_3d: two cases can be encountered:
(1) two hybrids pointing in opposite directions are equivalent to each other and their directions are perpendicular to the plane formed by the other three whose angles in relation to each other are 120° (hybridization $sp_3d_{z^2}$).
(2) Four equivalent hybrids form a square pyramid and the fifth orbital is perpendicular to the base (hybridization sp_3d_{xy}). In both cases, there are necessarily two different kinds of hybrid orbitals.

γ) sp_2d_{xy}: four coplanar orbitals of the same kind pointing in directions that are at 90° angles to each other.

δ) sp_3d_4: two cases can be considered:

Eight orbitals pointing to the apexes of a triangular dodecahedron with d_{yz}, d_{zx}, $d_{x^2-y^2}$ and d_{z^2} orbitals. The hybrids form two sets, each of four orbitals of the same kind.

Eight orbitals pointing to the apexes of a square antiprism with d_{xy}, d_{yz}, d_{zx} and $d_{x^2-y^2}$ orbitals. The eight orbitals are equivalent.

ε) $sp\,d_4$ (d_{xy}, d_{yz}, d_{zx}, $d_{x^2-y^2}$): six orbitals pointing to the apexes of a triangular prism.

e) Hybridization using f orbitals:

α) sp_3d_3f (d_{xy}, d_{yz}, d_{zx}, f_{xyz}): eight equivalent orbitals pointing to the apexes of a cube.

β) $sp_3d_5f_3$ ($f_{x(y^2-z^2)}$, $f_{y(z^2-x^2)}$, $f_{z(x^2-y^2)}$): twelve equivalent orbitals pointing to the middles of the edges of a cube.

In Figure 1, the directions of the orbitals described above are schematized and an example of molecule is given for each structure. All analytical expressions corresponding to these hybrid orbitals can be found in Appendix B.

4. Total Energy and Bond Energy

Although only total energy is meaningful, a more tractable form is obtained using localized molecular orbitals. Indeed, we can, with reasonable accuracy, give to the total energy (electron energy + repulsion energy between nuclei) of a molecule with "localizable" orbitals the general form [12],[13]:

$$E = \sum_{[\ell]} \varepsilon_\ell + \sum_{[ij]} \varepsilon_{ij} \qquad (14)$$

where ε_ℓ is the contribution from the nonbonding molecular orbitals φ_ℓ (i.e., corresponding to inner-shell atomic orbitals and eventually to lone pairs of classical schemes) and ε_{ij} is that from various orbitals φ_{ij}, which are localized between the nuclei $\underline{i}$ and $\underline{j}$ (corresponding to chemical bonds).

However, in the expression of the quantities ε_ℓ and ε_{ij}, we have two different kinds of terms. In the first kind, we only find terms coming from the orbitals themselves φ_{ij} (or φ_ℓ): kinetic energy,

9

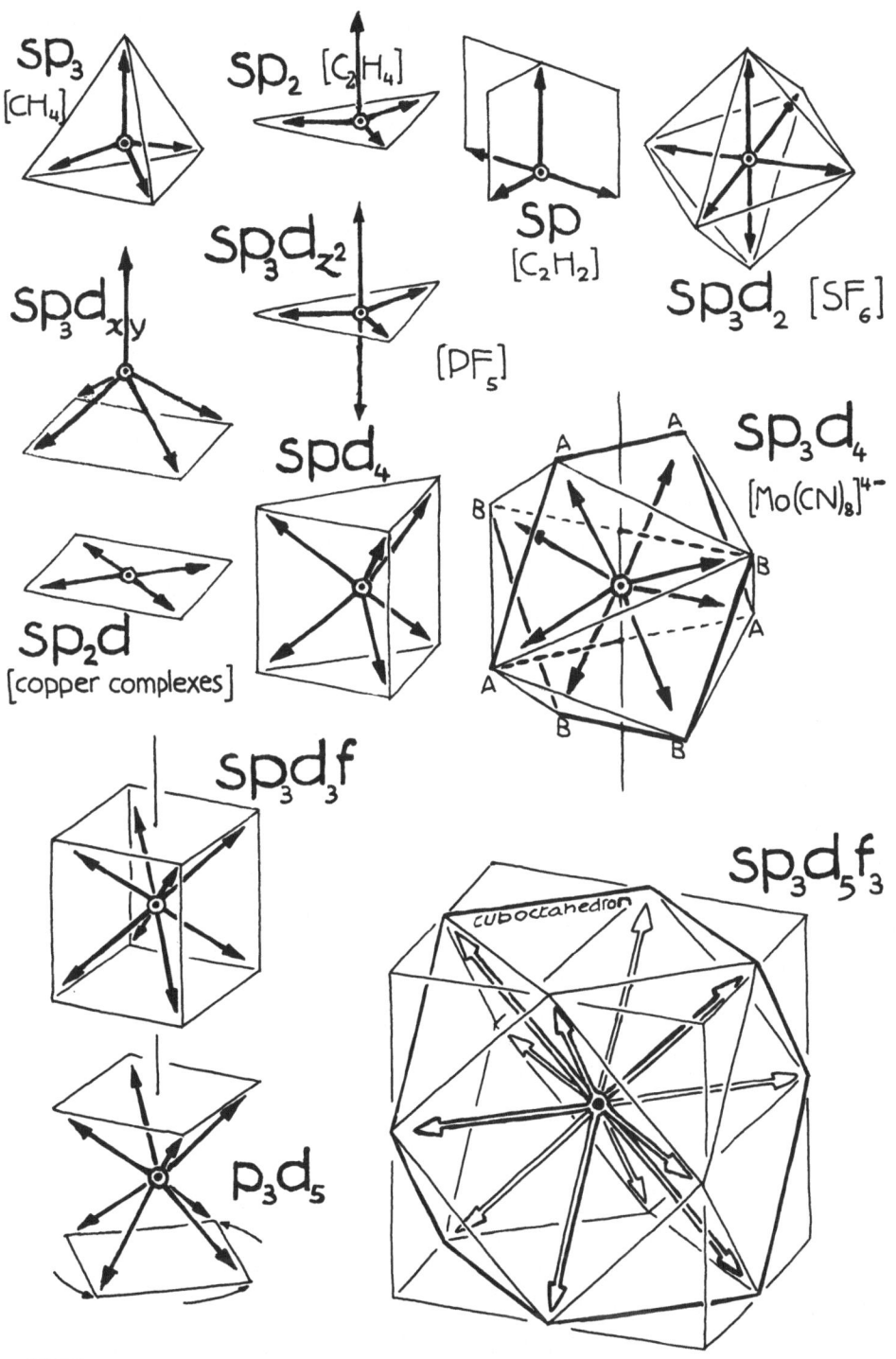

sp₃ [CH₄]

sp₂ [C₂H₄]

sp [C₂H₂]

sp₃d₂ [SF₆]

spd_{x,y}

sp₃d_{z²} [PF₅]

spd₄

sp₃d₄ [Mo(CN)₈]⁴⁻

sp₂d [copper complexes]

sp₃d₃f

sp₃d₅f₃

cuboctahedron

p₃d₅

⬚1 Main hybridization types in molecules

interaction energy with the nuclei i and j (or l). In the second kind, the terms correspond to the interaction between these orbitals on the one hand and the other orbitals and the nuclei different from i and j (or l) on the other. In fact, these last terms practically reduce to the electrostatic interaction between the density $\varphi_{ij}{}^2$ (or $\varphi_l{}^2$) and the net charges carried by the atoms different from i and j (or l). Consequently, the quantities ε_{ij} and ε_l, related to the various localized molecular orbitals, also depend on the environment of these orbitals. Therefore, they are not, in general, transferable from one molecule to another. However, in molecules in which net charges are never very important (as in organic molecules particularly), the quantities ε are practically independent of the molecule.

Rather than consider the total energy of a molecule, the chemist often prefers to consider the formation energy of the molecule from atoms. Therefore, this energy is the difference between the energy of the isolated atoms in their ground states and the total energy of the molecule. When using localized molecular orbitals one can divide this formation energy into contributions to the various localized molecular orbitals of the type φ_{ij}. If the quantities ε_{ij} are transferable and depend only on the atoms involved in the orbitals φ_{ij}, one can formally consider the energy of the various bonds and calculate its value. The increments, introduced by such systematics in order to obtain a better reproduction of the experimental values, are there in order to correct the modifications brought to the quantities ε by the molecular environment. However, in any case, regardless of the success obtained by the proposed systematics [14], we can never emphasize enough their arbitrary character.

5. Prediction of the Molecular Geometry

The geometry of a molecule is determined by the arrangement of all the nuclei, which minimizes the total energy. Therefore, the prediction of the geometry is essentially an energetic problem. However, we saw that, with a good approximation, one can consider a molecule in terms of more or less localized molecular orbitals, built up from atomic orbitals, usually hybrids. A localized molecular orbital, corresponding to a bond (in the classical sense), is built up from two hybrids pointing at one another or, as in CH_4, from one hybrid and the 1s orbital of a hydrogen. Its formation, as we saw in

the last section, lowers the energy of the system. Therefore, the
most stable structure will use the greatest number of basic hybrid
orbitals, in agreement, of course, with the number of electrons. For
example, in CH_4, the carbon atom has at its disposal the 1s orbital
and four sp_3 hybrids, and each hydrogen atom has one 1s orbital, for
a total of ten electrons. Therefore, we have one localized 1s
orbital and four bond orbitals. In NH_3, the number of available
orbitals for the nitrogen atom is the same as for the carbon atom, but
it has seven electrons instead of six. Therefore, we still have the
1s orbital of the nitrogen atom but only three localized molecular
orbitals corresponding to the N-H bonds. The fourth hybrid,
corresponding to the classical lone pair, is doubly utilized; it is
a dangling orbital.

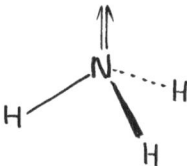

In a molecule such as ethane H_3C-CH_3, each carbon atom has four
sp_3 hybrids. Between the two carbon nuclei, there is a localized
molecular orbital built up from two hybrids, one for each carbon
atom, and pointing at one another. The alignment of the hybrid axes
gives the most stable system. Indeed the orbital overlap is maximum
and therefore the interaction between atomic orbitals is the
strongest: The corresponding energy ε_{ij} is the highest in absolute
value.

According to the hybridization concept, the building of a
molecule is similar to that of a child's construction set, i.e.,
the atoms fit into one another and the axes of their hybrid orbitals
are aligned. The restricted number of hybridization types not only
allows a clear classification of the various structures, [9] but also
permits their prediction. We shall see that the situation is the same
for crystals.

6. Fundamental Remarks Concerning the Alignment of Orbitals

In most molecules, bonding orbitals can be built up from two
hybrid orbitals pointing at one another along the nuclei line. In
fact, although this alignment corresponds to the most favorable

energetic condition, it is not absolutely necessary. Bonding orbitals can exist even if, for geometric reasons, orbital axes are not aligned. In this case, we have the so-called <u>bent</u> <u>bonds</u> (or <u>banana</u> <u>bonds</u>); e.g., cyclopropane [15,7]:

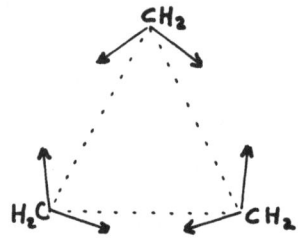

The angle between the two hybrids carried by each carbon atom is about 110°. The partial densities corresponding to each orbital are displaced to the outside.

A similar situation occurs in the CrO_8^{3-} ion. The chromium atom is surrounded by four O_2 groups, the oxygen atoms being sp-hybridized. An sp-hybrid of an oxygen atom is available to each hybrid orbital sp_3d_4 of chromium. Alignment of the sp orbitals with chromium hybrid orbitals is impossible to achieve.

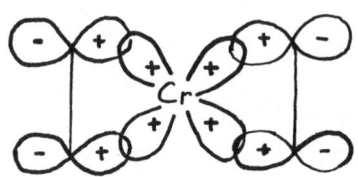

Therefore, the bends O-Cr are bent bonds. If we assume that the chromium orbital points toward the center of gravity of the positive lobe of the 2p orbital of the corresponding oxygen atom, the angle $\widehat{OCrO}$ must be smaller than the angle between the chromium hybrids. Experimentally, $\widehat{OCrO}$ is about 45°, the distance O-O: 1.4 Å and Cr-O about 1.9 Å [16,17]. The distance between the center of gravity of the positive region of the 2p orbital of oxygen atoms and the nucleus is about 0.3 Å. Therefore the angle between chromium hybrids is about 50°. According to the structure of sp_3d_4 hybrids (see Appendix B), this value is plausible.

The case of the $Ti(NO_3)_4$ molecule is similar; the nitrate groups are fixed on the titanium atom like bridges by two oxygen atoms [17].

In molecules, bent orbitals are rare. However, in crystals, in which the lattice periodicity imposes severe constraints, this phenomenon is much more frequent.

7. Electron Charges and Bond Polarities

In every point of space, the electron density is given by the square of the total wave function modulus: $|\psi|^2$. Maps can be drawn. However, the chemist prefers to know the average amount of electricity around each nucleus. The partitioning of the total density into contributions from various atomic orbitals allows a definition of the orbital populations and of the electron charge carried by various atoms in the molecule [11]:

$$q_A = 2 \sum_i \sum_{r \in A} (c_{ir}^2 + \sum_{s \neq r} c_{ir} c_{is} S_{rs}) \qquad (15)$$

Indeed, the steep decrease of atomic orbitals as a function of the distance to the nucleus permits, practically speaking, identification of the total charge around the nuclei with the sum of the corresponding atomic populations.

The net charge is defined as the difference between the nuclear charge N and the total electron charge around it:

$$Q_A = N_A - q_A \qquad (16)$$

In general, since atoms are not necessarily of the same kind, the center of gravity of the electron cloud do not coincide with that of the positive charge of the nuclei. The molecule has a dipole moment $\vec{\mu}$. Classically, this moment is partitioned into contributions $\vec{\mu}_i$ from each bond or lone pair [7,19]:

$$\vec{\mu} = \sum_i \vec{\mu}_i \qquad (17)$$

Since electrons are not localized in the molecule, the various partial moments $\vec{\mu}_i$ have, of course, no physical meaning. However, using the description in terms of localized orbitals, these moments $\vec{\mu}_i$ are easily calculated from the centers of gravity of the corresponding partial densities $|\varphi_i|^2$.

According to the electronegativity of orbitals [7], which the functions φ_i are built upon, the densities $|\varphi_i|^2$ will be more or less displaced toward one or the other nucleus. Conventionally, the bond is said to be more or less polar. According to even older terminology, a bond is said to be <u>covalent</u> when the density $|\varphi_i|^2$ is symmetrically distributed between the two basic orbitals, as for example in the H_2 molecule or between the two carbon atoms in ethane, H_3C-CH_3. Otherwise, and this is the general case, it is said to be partially <u>ionic</u>. Of course, a completely ionic bond cannot exist in a molecule since it would be dissociated into two ions. Net charges carried by atoms in a neutral molecule do not exceed in general some tenth of an electron [19].

II. Application to Crystals

1. The Two Large Classes of Crystals

Considering the crystal as a giant molecule, one could suppose that after having performed an self-consistent field (SCF) treatment, which yields delocalized molecular orbitals belonging to the group of lattice symmetry and therefore possessing translational network symmetry, one can then localize these orbitals by suitable unitary transformation. As with the molecules of finite size, one would obtain: (1) localized orbitals around diverse nuclei, identifying themselves with the inner atomic orbitals of isolated atoms, (2) localized orbitals around one or more nuclei, corresponding to lone pairs or to bonds, in the classical sense, and (3) eventually also molecular orbitals that are not localizable.

A new situation, however, appears in crystals. Experiments show that, in effect, there are two large classes of crystals: (1) those in which the nuclei are grouped in the form of small, identical, periodically arranged islets, and (2) those in which the nuclei are not distributed in this manner. For example, in the benzene crystal the islets consist of six nuclei of carbon and six of hydrogen, whereas in a crystal of sodium chloride, the repartition of sodium and chlorine nuclei is homogeneous.

This situation results in crystals of the first type the molecular orbitals can be localized around each of the islets. On the other hand, as shown in a simple case in Appendix C, the molecular orbitals localized around each islet are practically identical to those that would be obtained for isolated molecules corresponding to the group of nuclei of each islet; thus one can consider the crystal as being formed by the packing of molecules of finite size; e.g., C_6H_6 in the benzene crystal, P_4 in white phosphorus, S_8 in sulfur, etc. Such crystals are called <u>molecular crystals</u>.

When the nuclei are arranged regularly in space without leaving a vacancy, the crystal can be represented by a giant molecule (<u>macromolecular crystal</u>) and, as in molecules, the molecular orbitals would be more or less localized, which step by step assure the bonds. Two typical cases are possible here. The network of the bonding orbitals could extend itself either in three spacial directions or in only two or one direction. In the first case, one would speak of a <u>three-dimensional</u> macromolecular crystal, as a true generalization of a molecule of finite size, while in the other case, one would have either a two-dimensional lamellar network or a monodimensional fibrous network. The crystal would therefore appear as being intermediate between the molecular type (in the proper sense of the term) and the three-dimensional macromolecular type, because it is formed by the layering of distinct macromolecules in one or two dimensions. Examples: linear chains with Te, Se ; layers with graphite or talc.

2. Molecular Crystals

The total wave function can be constructed on molecular orbitals that are almost perfectly localized around each of the islets of the nuclei. Furthermore, these orbitals are practically identical to the orbitals of small-sized molecular entities, which can be considered to correspond to islets assumed to be isolated from one other.

By definition, the cohesive energy in such a crystal is the difference between the actual energy of the crystal and the sum of the energies that corresponding isolated molecules in space, would have.

Although the total energy of a crystal has only a physical signification, one can consider it as the sum of two terms: (1) energy of the ensemble of isolated and nondeformed molecules, and (2) energy of interaction [11]. In the second term, which consequently corresponds to the cohesive energy, electrostatic interactions between molecules play a role: if the molecules are neutral, dipolar interaction (as the molecules are polar), Van der Waals interaction, whose contribution in magnitude is inferior to that of dipolar interaction, etc. The associations called hydrogen bonds, which exist between certain groups A-H and a molecule carrying a lone pair in the classical description [7], can be attached to the electrostatic interaction:

$$A-H...B$$

In effect, the greater part of the energy of interaction is due to the electrostatic interaction between A-H and B groups.

Because of its origin, the cohesive energy of a molecular crystal is always weak and is of the order of a few kcal/mol. For this reason, a small external energy can easily displace (or even disperse) the diverse entities of the crystal with respect to one another. A molecular crystal would always have low hardness and low fusion or sublimation points. If the molecules constituting the crystal are not polar, only Van der Waals forces are responsible for cohesion and in such a case, the melting-point would be lowered in direct proportion to the decreasing size of the molecules or the molecular weights (Trouton's Law: ML/T = Cte). In the presence of dipolar interactions or in the presence of hydrogen bonds (ice), the melting-point is appreciably elevated. In crystals with lamellar structure the different layers can easily glide over one another. The crystal would be unctuous to the touch and would possess lubricating properties (graphite, talc, molybdenum sulphide, etc.). However, with increasing size of constituent elements, the fusion point of the crystal is considerably elevated. The same is true for a crystal of fibrous structure (tellurium). The mineralogic hardness values (in Mohs scale, see Appendix D) and the temperatures of fusion or sublimation of some molecular crystals and the corresponding type of preponderant interaction are indicated in Table 1.

It is known that in a molecular crystal, the entities begin to turn about their own axes, when the temperature increases. This rotation, which occurs long before the melting point, explains the discontinuity in specific heat and the narrowing of the NMR lines (e.g., in benzene at 110 K). This fact is not incompatible with a description of crystal as a giant molecule. In a small molecule (e.g., ethane), the molecular orbital localization does not forbid an internal rotation. The localized molecular orbitals follow the nuclei in their rotation. Likewise, in molecular crystal, when the temperature increases, various localized molecular orbital sets can rotate, without dislocation for the crystal.

Finally, we shall note that the conformation of various entities is not necessarily the conformation which is the most stable in vacuo. This phenomenon is due to the fact that the crystal field variations are often greater than the potential barriers of rotation [18].

2 Molecular crystals

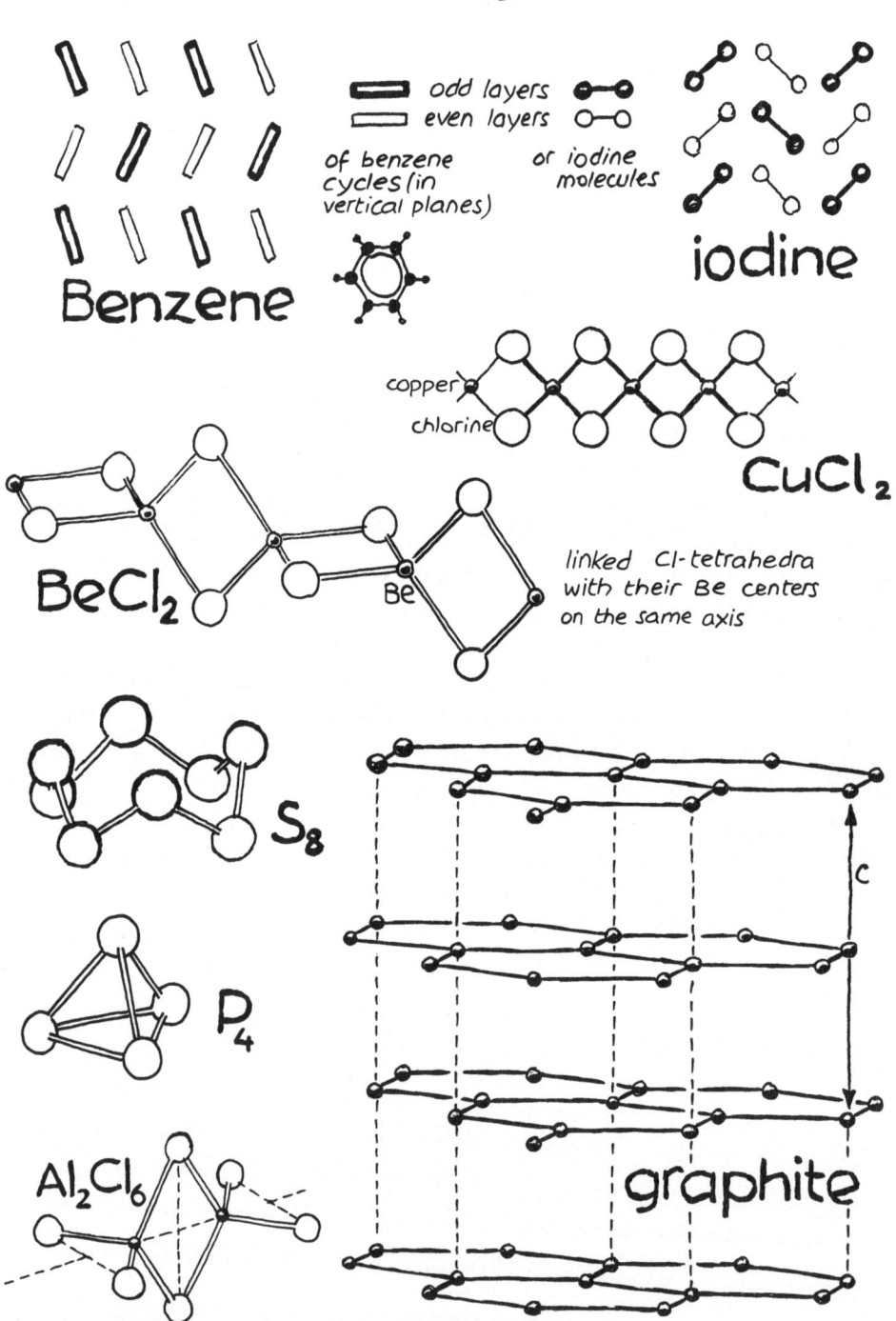

odd layers
even layers

of benzene
cycles (in
vertical planes)

or iodine
molecules

Benzene

iodine

copper
chlorine

CuCl₂

BeCl₂   Be

linked Cl-tetrahedra
with their Be centers
on the same axis

S_8

P_4

Al_2Cl_6

c

graphite

Table 1

Crystal	Hardness	Melting or Sublimation (*) point (°C)	Preponderant type of interaction
Helium	–	-271	Van der Waals
Neon	–	-249	"
Oxygen	–	-218	"
Nitrogen	–	-210	"
Sulfur	2	119	"
White phosphorus	0.5	44	"
Tellurium	2.5	452	"
Iodine	–	114	"
CO_2	–	-79*	"
Ice	∿2	0	H-bond
PF_5	–	-94	Van der Waals
P_4O_6	–	23	"
P_4O_{10}	–	300*	dipolar
$A\ell_2C\ell_6$	–	194	Van der Waals
SnI_4	–	145	"
SiF_4	–	-90	"
CH_4	–	-184	"
$C\ell CH_3$	–	-98	dipolar
C_6H_6 (benzene)	< 1	5.5	π
C_2H_5OH	< 1	-117	H-bond
Sucrose	2.5	–	H-bond
Graphite	2	>3500*	π

Since the stacking of constituent elements in a molecular crystal requires weak energy, various factors come into play, so it is difficult to give simple general rules for predicting which crystal structure will be adopted. In two cases, however, this is possible. If the crystal is formed by the stacking of atoms (rare gases), the stacking would be as compact as possible, e.g., a cubic close-packed structure in general (except for the helium crystal, which has a hexagonal close-packed structure). The presence of hydrogen bonds, owing to their directional nature (alignment of A-H and B groups), imposes a network geometry (ice). In the case of rare gases, the cohesive energy is, in general, calculated using empirical potentials: Lennard-Jones or Mie potentials [20]. For ice, more elaborate calculations using the molecular orbitals method have been published [21]. The hexagonal structure is somewhat more stable than the cubic structure.

For molecular crystals the following examples can be cited:

- rare gases, metalloids (H_2, O_2, Cl_2, I_2, S_8, P_4, etc):
    In these, cohesion is due to Van der Waals forces.

- light mineral compounds (CO_2, SO_2, nitrogen oxides, H_2O), halides (Al_2Cl_6, Al_2Br_6, SnI_4, PF_5), $Ni(NO_3)_4$, ...

- nearly all the organic compounds (except the metallic salts of organic acids) viz. saturated or unsaturated hydrocarbons, alcohols, amines, acids ...

    In hydrocarbons the cohesion is of the Van der Waals type. In others the cohesion is due to dipolar interactions or to hydrogen bonds.

Figure 2 shows arrangements in various molecular crystals.

Within the group of molecular crystals, one can include the crystals formed from edifices of a different nature, in general carrying electron charges. These can be described in a manner similar to that used for the entities in the molecular crystals discussed above, by localized orbitals around diverse groups of nuclei.

Consider, for example, phosphorus pentachloride. Classically, the PCl_5 crystal can be written as being composed of PCl_4^+ and PCl_6^- ions. Each chlorine atom has, on the one hand, a utilized atomic orbital for forming a bonding molecular orbital with one orbital of phosphorus, and, on the other hand, three nonbonding, doubly utilized atomic orbitals. All of the orbital systems that one could imagine as constructed on the peripheral orbitals of chlorine, will necessarily comprise as many bonding orbitals as antibonding orbitals. All the orbitals are utilized twice. The electron energy would be the same as that of isolated atomic orbitals. In other words, there would be no stabilization with respect to isolated PCl_4^+ and PCl_6^- ions. One would say that there is no bonding. In addition, since all the atomic orbitals are used twice, no electron transfer from one group to other is possible (i.e., when one limits himself to the minimal basis of atomic orbitals for describing the system). Cohesion is assured simply by electrostatic attraction between the ions. Short-range forces prevent the collapse of the system on itself.

The situation is similar in pentabromide. However, since PBr_6^- is unstable, the crystal would be formed from PBr_4^+ and Br^- ions. Br^- ions carry only doubly utilized atomic orbitals, which behave in a similar manner as atoms of chlorine in PCl_6.

Other examples of crystal in this category include: ammonium salts, $(NH_4)Cl$ f.ex. .

We shall see further that the situation is completely different in alkali halides.

3. Three-dimensional Macromolecular Crystals

The edifice can be considered as a giant molecule. It could therefore be described in terms of more or less localized molecular orbitals formed from suitable atomic orbitals, hybrid or not. As in molecules, one would have localized orbitals identical to inner-shell atomic orbitals. In addition to the localized orbitals between two nuclei, corresponding to the classical straight line between atoms that indicates a bond, there could be atomic orbitals, hybrid in general, which are not engaged in molecular orbitals localized between couples of nuclei. These could be either doubly utilized (nonbonding orbitals), corresponding to lone pairs, or unutilized, corresponding to the classical vacancies (viz. $2p_z$ orbital in BF_3). Again, more frequently than in the molecules, one would have delocalized systems constructed from a finite number, usually very limited, of atomic orbitals. Finally, one also comes across systems possessing an orbital system completely delocalized over the whole crystal.

In molecules, the presence of atomic orbitals that are not used in forming the bond, causes a decrease in the number of linked atoms. Evidently, in the case of crystals, the same would also be true, but the decrease in the number of linked atoms would make the construction of a three-dimensional network impossible. One would therefore have giant molecules forming either layers or chains.

Because the different types of possible hybridizations are limited, as in the case of molecules [11], it is easy to classify the different structures that could appear. Although the lamellar and fibrous macromolecular crystals have been included in the previous paragraph, they will be a second time mentioned for logic in classification.

The various structures are arranged in the following categories:

1) Crystals permitting a description in terms of localized orbitals,
2) Crystals presenting delocalized systems of finite size,
3) Crystals possessing an infinite delocalized system.

It should be noted that the classification proposed here is essentially qualitative. In fact, it has been observed that in molecules, angular deformations with respect to the most symmetric type of hybridization, are always weak. Therefore the possible effects of distortions that may appear will be neglected. The hybrid orbitals, and consequently the corresponding localized molecular orbitals are, in any case, less influenced by these effects. In this section, the existence of bent bonds will only be mentioned. A more complete study is included in Part IV.

On the other hand, in general, no indication will be given of the crystal system to which the substance belongs. This problem will be discussed later on.

1) Crystals permitting a description in terms of localized orbitals:

The orbitals corresponding to the inner electron shells of atoms will not be considered because they do not occur in the structure of a crystal. Although several hybridization combinations are possible, only the principal ones will be described here, with a few typical examples. The structures are derived from reference [16].

a) Compounds of the formula A_mB_n where A is a metal and B is a nonmetal

α : $sp_3 \times sp_3$

The atoms A and B, sp_3-hybridized, each lie at the center of a tetrahedron. Depending upon the atoms carrying or not carrying utilized or unutilized nonbonding orbitals, various structures are possible.

• Without nonbonding orbitals: two arrangements are possible, depending upon whether the network is formed uniquely from cycles of six atoms in form <u>chair</u> (<u>zinc-blende</u>), or mixture of <u>chair</u> and <u>boat</u> forms (<u>wurtzite</u>).

→ zinc-blende structure: diamond, silicon, germanium, BN, GaAs, AℓAs, CuCℓ, CuI, ZnS (zinc-blende),

→ wurtzite structure: ZnS (wurtzite), CdS, GaN, AℓN, BeO, ZnO.

• One of the atoms has two nonbonding utilized orbitals:

→ quartz (SiO_2), GeO_2 : three-dimensional network,

→ SiS_2, $TℓS_2$, $BeCℓ_2$, $MgCℓ_2$: infinite chains.

• One of the atoms has one utilized nonbonding orbital and the other atom has two utilized nonbonding orbitals:

→ As_2O_3, Sb_2O_3, PbO: layers

• The atoms are of same nature and carry a doubly utilized nonbonding orbital:

→ Black phosphorus,

→ As, Bi, Sb : layers

β : $sp_2 \times sp_2$

One has plane layers: graphite, BN. Each atom carries one <u>p</u> unhybridized atomic orbital, entering in a π-delocalized conjugated system over the entire layer.

γ : $sp_3 \times sp$

The sp-hybridized atom carries, in general, doubly utilized nonbonding orbitals.

→ Cristobalite (SiO_2), tridymite (SiO_2) derived, respectively, from a network of silicon atoms, zinc-blende, or wurtzite, with one atom of oxygen situated in the middle of each Si-Si segment; BeF_2 of cristobalite structure.

→ Cu_2O (cuprite), Ag_2O with sp_3-hybridized oxygen atom and sp-hybridized Cu or Ag atom. The structure is of inversed cristobalite.

δ : $sp_3d_2 \times sp_3$

The metal is linked to six neighbors and is at the center of an almost regular octahedron. The nonmetal, linked to four metallic atoms, is at the center of a tetrahedron.

→ One nonbonding orbital on the metal:

SnS_2, CdI_2, $CdCℓ_2$: layers

→ Two nonbonding orbitals or nonmetal:

$CrCℓ_3$ in layers

→ Pb_3O_4 (red lead): two types of lead atoms, those sp_3d_2-hybridized and those sp_3-hybridized, joined by sp_3-hybridized oxygen atoms with two or one nonbonding orbital.

→ Fe_3O_4 (magnetite), Co_3O_4: spinel structure (see b-ε p.27).

→ Al_2O_3 (corundum), Fe_2O_3 (hematite), Cr_2O_3, $FeTiO_3$ (ilmenite) where one of every two atoms of iron in hematite is replaced by a titanium atom.

The metal is sp_3d_2 hybridized. The oxygen atoms are sp_3 hybridized. The bonds are bent. The structure will be studied in detail p.82.

ε: $sp_3d_2 \times sp_2$

The nonmetal is linked to three metal atoms and carries one doubly utilized bonding orbital:

→ TiO_2 (rutile), GeO_2, SnO_2, PbO_2, MgF_2, NiF_2, CoF_2, FeF_2, MnF_2, ZnF_2, the bonds are bent (<u>vide infra</u>).

ζ: $sp_3d_2 \times sp$

→ The nonmetal is aligned with two atoms of metal and carries two doubly utilized nonbonding orbitals: CrO_3, WO_3, ReO_3, ScF_3, AlF_3, FeF_3, CoF_3, PdF_3.

→ UF_5 in parallel chains formed from atoms of uranium sp_3 hybrid, with two nonbonding orbitals and carrying in addition four fluorine atoms sp-hybridized with three nonbonding orbitals.

η: $sp_3d_2 \times spd_4$

NiAs, FeS, CoS: the metal, sp_3-hybridized, is at the center of an octahedron and the nonmetal, sp_3d_4-hybridized, is at the center of a triangular prism.

θ: $sp_3d \times sp_3$

The metal is linked to four nonmetals and is in their plane:

→ CuO (tenorite), PtO, PdO, PtS (cooperite): three-dimensional network. The bonds are bent (<u>vide infra</u>).

→ $CuCl_2$, $CuBr_2$, $PdCl_2$: layers

ι: $sp_3d \times sp_2$ ($d = d_{z^2}$)

→ LaF_3, CeF_3, AcF_3, UF_3: layers

κ: $sp_3d \times sp_3$ ($d = d_{xy}$)

→ PbO: layers, constituted of PbO_4 pyramids having their bases in common. The lead atoms are alternatively over and under the plane of the base. Each atom of lead carries one doubly utilized nonbonding orbital.

λ: $sp_3d_3f \times sp_3$

→ The metal is surrounded by eight nonmetal atoms situated on the corners of a cube. The nonmetal is linked to four metallic atoms (<u>fluorite</u> structure):
CaF_2 (fluorite), SrF_2, BaF_2, HgF_2, PbF_2, UO_2.
→ The roles of metal and nonmetal atoms are reversed (<u>antifluorite</u> structure):
alkali oxides, sulfides or tellurides (LiO_2...) $SiMg_2$, $GeMg_2$, $SnMg_2$, $PbMg_2$, CBe_2.
→ In the fluorite structure, the metal atom carries two doubly utilized nonbonding orbitals. As a result it is linked to only six atoms of nonmetal:
Mn_2O_3, rare-earth metal oxides (<u>C</u>-structure).
→ In antifluorite structure, if the metal atom has two utilized nonbonding orbitals, it is linked to only six neighbors: Bi_2O_3, Zn_3P_2, which has a slightly distorted network.

μ: Other hybridizations permitting seven ligands
→ Monoclinic ZrO_2 (baddeleyite), HfO_2. The seven oxygen atoms form a polyhedron with one tetragonal base and one trigonal base.
→ Sesquioxides of La, Ce, Pr, Nd (<u>A</u>-M_2O_3 structure). The seven oxygen atoms form an octahedron with an additional oxygen atom above one of the octahedron faces.

sp_3d_3 hybridizations are possible. Perhaps <u>f</u> orbitals also intervene. These hybridizations are unstable. At high temperatures ZrO_2 has the rutile structure, and La_2O_3 and Nd_2O_3 crystallize with the <u>C</u>-M_2O_3 structure.

b) Compounds of A_mB_n formula where <u>B</u> is a complex structure

This category includes those crystals in which <u>B</u> is a polynuclear structure described by localized orbitals, such as oxysalts (sulfates, silicates, ...). The atoms situated on the periphery of $\underline{B}$ structure carry the atomic orbitals pointing in sufficiently open directions to be able to form localized molecular orbitals with orbitals of the metal atoms.

These crystals are arranged according to the hybridization of metallic atoms $\underline{A}$, indicating (eventually) the structure of the $\underline{B}$ group. The $\underline{A}$ atoms may not all be of the same nature.

α: sp_3

The metal atoms form four localized molecular orbitals with four neighboring atoms

      $\rightarrow$ Li_2SO_4 (t < 575°C), Ca_2SO_4 (t > 660°C),

        Be_2SiO_4 (phenacite), Zn_2SiO_4 (willemite), Li_2WO_4,

        Li_2MoO_4,

      $\rightarrow$ $LiA\ell SiO_4$ (eukryptite), $LiA\ell GeO_4$, $LiZnVO_4$.

In these crystals, the $\underline{B}$ groups (sulphate, silicate, tungstate, molybdate, etc.) have a tetrahedral structure and the central atom is sp_3-hybridized. Each peripheral oxygen atom carries three orbitals available for forming one or two molecular orbitals with the metal atom $\underline{A}$, which necessarily has one or two nonbonding orbitals.

β: sp_3d_2

      $\rightarrow$ Mg_2SiO_4 (forsterite), Fe_2SiO_4 (fayalite),

        $(Mg_{1.8} Fe_{0.2})$ SiO_4 (olivine), $A\ell_2BeO_4$ (chrysoberyl),

        $CrVO_4$, $CuCrO_4$.

      $\rightarrow$ MSO_4 where M = Cu, Mg, Co, Zn. $\underline{M}$ is at the center of a distorted octahedron. The SO_4 groups are deformed $(r_{SO} = 1.48$ at 1.53 Å).

      $\rightarrow$ FeS_2 (pyrite). The sulfur atoms are sp_3-hybridized and each atom of sulfur is engaged, on the one hand, with another sulfur, and on the other hand, with three atoms of iron. Each iron atom, sp_3d_2-hybridized, forms localized orbitals with six neighboring sulfur atoms.

      $\rightarrow$ $Na_3A\ell F_6$ (cryolite). The sodium atom is linked to six fluorine atoms carried by different $A\ell F_6$ octahedral groups. Four Na-F bonds are bent (_vide infra_).

      $\rightarrow$ $NaNO_3$, $LiNO_3$, $CaCO_3$ (calcite), $MgCO_3$, $FeCO_3$ (siderite), $ScBO_3$, $InBO_3$, YBO_3. The M-O bonds are bent (described later).

γ: sp_3d

→ Pd $(CN)_2$. One has plane layers, in which the CN groups establish linear bridges between Pd atoms situated at the nodes of a square lattice.

δ: $sp_3d_5f_3$

→ $CaTiO_3$ (perovskite), $SrTiO_3$, $PbZrO_3$, $KMgF_3$.
The titanium, zirconium, and magnesium atoms are sp_3d_2-hybridized. The oxygen atoms are situated at the apexes of regular octahedrons connected by sp-hybridized oxygen atoms.

ε: sp_3 and sp_3d_2

→ The spinels: $MgA\ell_2O_4$ (ruby spinel) in which the Mg (or Fe) atoms are sp_3d_2-hybridized and $A\ell$ (or Cr) atoms sp_3-hybridized.

→ $Be_3A\ell_2(Si_6O_{18})$ (beryl) constituted by Si_6O_{18} rings, bound to each other by sp_3-hybridized Be atoms and sp_3d_2- -hybridized $A\ell$ atoms. The metal-oxygen bonds are probably weakly bent.

ζ: sp and sp_3d_4f

→ $K_2PtC\ell_4$, $K_2PdC\ell_4$. The potassium atoms carry eight hybrid orbitals which point toward the apexes of a square parallelepiped. The chlorine atoms are sp-hybridized. The alignment of orbital axes is not possible. The bonds are bent. This structure will be studied in more detail in Part IV.

η: sp_3d_4

→ $ZrSiO_4$ (zircon), $CaCrO_4$, $CaWO_4$ (scheelite), $NaC\ell O_4$; MPO_4, $MAsO_4$, MVO_4, where $\underline{M}$ is a lanthanide ($ScPO_4$, YPO_4, $ThPO_4$...).
The metal is surrounded by eight oxygen atoms. We shall study this fundamental structure with bent bonds in Part IV.

2) Crystals presenting delocalized systems of finite size:

Such crystals generalize the exceptional cases mentioned above in small molecules: boron hybrids, ferrocene.

We shall distinguish three cases:

a) crystals composed of elementary cells, which are composed of finite delocalized systems (cellular structure),

b) crystals composed of a rigid network of localized orbitals, in which the atoms or molecules locally form delocalized orbitals with certain atoms of the lattice (cage structure).

c) crystals composed of delocalized systems binding localized systems (mixed structure)

a) Cellular structure
....

To explain this type of structure [22], we have chosen sodium chloride. The sodium atom has one valence electron and the chlorine atom, seven electrons. Each atom is surrounded by six neighbors of the other element. Consequently, in classical theory, it is impossible, with the eight available electrons, to build two electron bonds between the sodium and chlorine atoms and still respect the equivalence between the neighboring atoms. The situation is the same for localized molecular orbitals, doubly utilized. In return, it is possible to describe the system by means of four delocalized molecular orbitals, constructed on the s and p orbitals of each chlorine atom, as well as on the sp_3d_2 hybrid orbitals of the neighboring sodium atoms pointing toward the chlorine atom. The delocalized system is constructed from ten atomic orbitals (hybrid or not). This structure is a generalization of the case of B_2H_6, where each of two molecular orbitals is constructed from three atomic orbitals: two hybrids of boron and the 1s orbital of the bridged hydrogen.

The sodium chloride crystal is described by the ensemble of these cubic cells, which fill the entire space.

In cesium chloride, CsCl, where each atom is surrounded by eight equivalent neighbors, the orbitals of the elementary cells are built from the four s and p orbitals of chlorine atom, as well as from the eight sp_3d_3f hybrid orbitals of the neighboring cesium atoms, pointing toward the chlorine atom.

Examples:

→ Rock-salt structure: The all-alkali halides at ordinary temperatures [except CsCl (if t < 445°C), CsBr and CsI]; MgO, CaO, SrO, BaO, MnO, FeO, NiO, CoO, CdO, VO, UO ; MgS, CaS, BaS, PbS, MnS.

→ CsCl structure: CsCl, CsBr, CsI, TlCl ; RbCl, RbBr and RbI under high pressure (> 5000 kg/cm^2); RbCl at high temperature.

→ CaC_2 structure: the crystal is made from chains,

... Ca-C$_2$-Ca-C$_2$-Ca ... , composed of sp$_3$d$_2$-hybridized calciums, linked to C$_2$ groups by localized orbitals. The chains are linked in twos by delocalized systems of finite size, constructed with the π and π' orbitals of each C$_2$ group and four calcium atoms situated in the mediator plane of C$_2$.

b) Cage structure

There are molecular orbitals constructed from atomic orbitals, hybrid in general, belonging to certain atoms of a tridimensional rigid framework, and from orbitals of one or many metallic atoms.

This structure is very usual, particularly in natural silicates. We shall give only some examples:

→ Tectosilicates and feldspars: If in the tridimensional framework corresponding to the stoichiometric formula SiO$_2$ (e.g., tridymite or cristobalite) a certain proportion of silicium atoms is substituted by aluminum atoms, in order for the lattice to preserve its structure, it is necessary that the crystal acquire one supplementary electron per aluminum atom. In order to restore the electroneutrality, the crystals accept ions, e.g., Na$^+$, which penetrate the holes of the lattice and become linked to the oxygen atoms.

E.g.:
- Cristobalite framework in which one-third of the silicium atoms are substituted by aluminum atoms: K(AℓSi$_2$O$_6$) (leucite).
- Tridymite framework, where half of silicium atoms are substituted by aluminum atoms: KNa$_3$(AℓSiO$_4$)(nepheline). The sodium atoms are linked to seven oxygen atoms and the potassium atoms to nine oxygen atoms.
- Other SiO$_2$ framework:
  Feldspars: K(AℓSi$_3$O$_8$)(orthoclase),
  Na(AℓSi$_3$O$_8$)(albite), Ca(Aℓ$_2$Si$_2$O$_8$)(anorthite).
  These structures correspond to delocalized systems of finite size, similar to the ferrocene system.
- CaB$_6$: The calcium atoms are situated in the center of the holes of the tridimensional framework of an octahedra of boron atoms, situated in nodes of a simple cubic lattice and joined up by their corners.

To this type of structure, we shall connect the important group of zeolites. These minerals are tectosilicates, which possess large channels in which water molecules are weakly bound by hydrogen bonds. Located in surroundings, with altered hygrometry, the mineral reversibly loses and takes up water.

The ejected water molecules can be replaced by various small molecules: SH_2, NH_3, CS_2, $CC\ell_4$ and even by mercury or helium atoms.

In addition, the zeolites contain metal atoms, linked to a $A\ell_x Si_y O_z$ framework by delocalized systems of finite size. These metal atoms are weakly linked and can be interchanged with other metal atoms. These interchanges are reversible and respect the electroneutrality of the crystal:

$$Ba^{2+} \rightleftharpoons 2\,K^+ \quad \text{or} \quad Ca^{2+} \rightleftharpoons 2\,Na^+$$

The property is utilized in artificial resins (permutites).

In zeolites, the H_2O molecules are linked to oxygen atoms of the framework, and to metallic ions. The hydrogen atoms form hydrogen bonds with two oxygen atoms of the framework, and the two nonbonding hybrid orbitals carried by oxygen are utilized to build a delocalized system with metal (e.g., Na) and other nonbonding orbitals of oxygen atoms of the framework.

c) Mixed structure

Included in this category are crystals in which the metal atom is situated in the center of a more or less delocalized system of finite size (as in cage structure); in addition, however, the metal atom forms a bridge between localized islets. In other words, the situation is similar to that described in 1-b α to δ, but the orbital system centered on the metal cannot be described by localized orbitals. In general, the metal atom is surrounded by a large number of neighbors (oxygen atoms): 7, 8, 9, 11, 12. Though the M–O distances are very similar, the spatial distribution of oxygen atoms around the metal is very disconcerting. Consequently, it is difficult to introduce suitable hybrid orbitals to describe the system in terms of localized orbitals. However, it is possible that a more detailed analysis of the structure of compounds included in this category would indicate the possibility of obtaining such hybrids. Some modification of our classification may be necessary.

This structure is usual in the salts. The following examples indicate the diversity of cases:

→ $BaSO_4$, $BaCrO_4$, $BaBeF_4$, $PbSeO_4$: the M-O distances are similar.

→ Àxinite (complex borosilicates): the calcium atom has ten oxygen atoms as neighbors.

→ $RbNO_3$, $CsNO_3$: the metal has twelve oxygen atoms as neighbors.

→ $M_xV_2O_5$, where M is any metal [23]. The atoms M are situated in cages composed of the oxygen atoms. The positions of the metal atoms are very unexpected. Only calculation of the energy should yield an explanation of these privileged positions.

3) Crystals possessing an infinite delocalized system:

Two cases are to be distinguished:

a) The localized orbitals correspond only to inner atomic orbitals and the crystals possess a completely delocalized orbital system: we have a metal.

b) The crystal possesses localized orbitals (other than the inner orbitals) constituting layers between which a delocalized system is inserted.

→ $CaSi_2$: the silicon atoms arranged in cycles of six atoms (form chair) form layers. The delocalized system is built on the sp_3 axial orbitals of the silicon atoms and on the atomic orbitals of calcium atoms situated between the layers. Each calcium is surrounded by six silicons.

→ AlB_2: the structure is similar, but the boron layers are planar and possess the graphite structure.

This swift survey of principal structures of crystals clearly shows that, for a qualitative description, no one difference appears between the molecules of finite size and the crystals. It should possible to explain the structure of glass by introducing a disorder in the lattice.

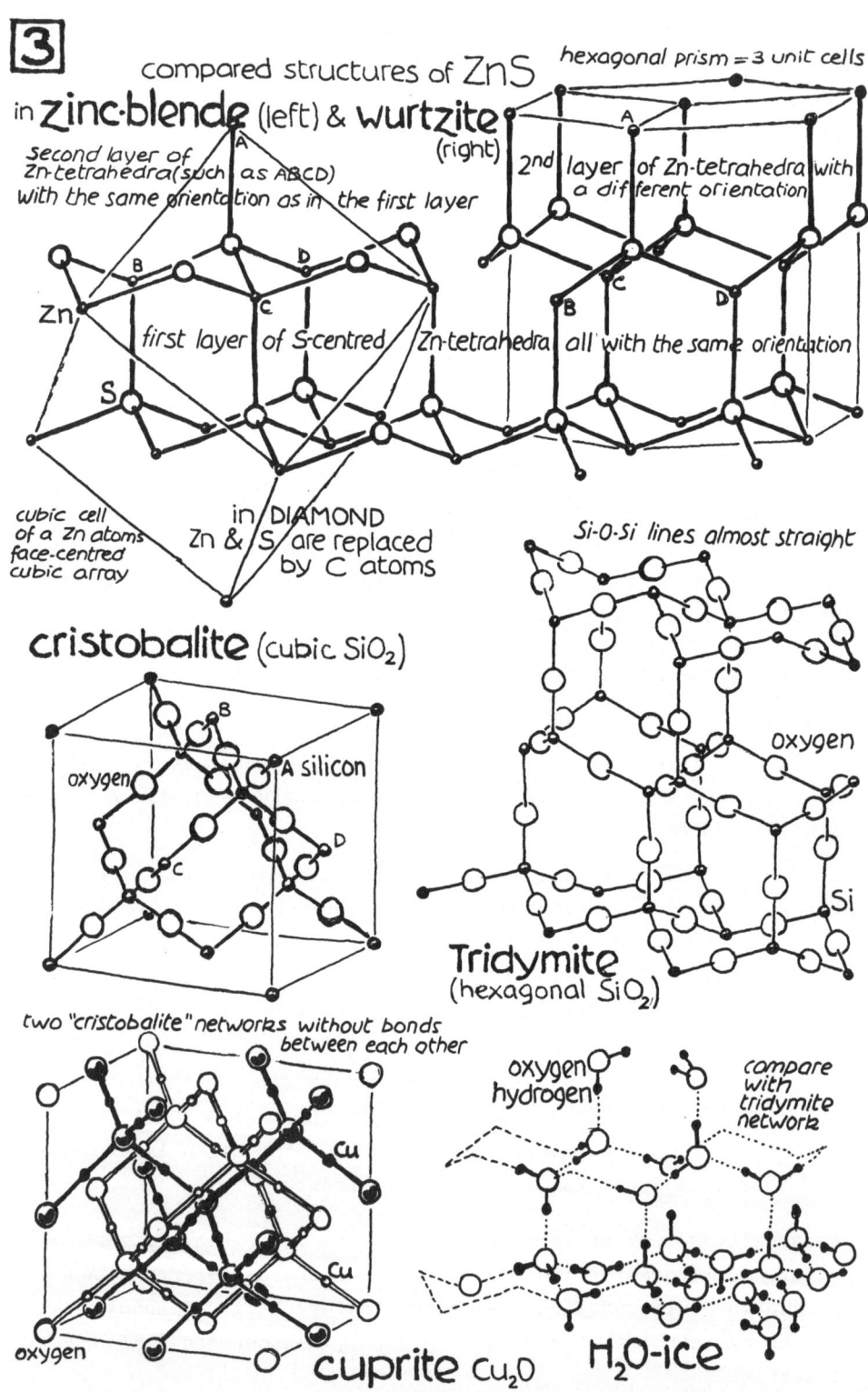

3 compared structures of ZnS
hexagonal prism = 3 unit cells
in **zinc-blende** (left) & **wurtzite** (right)

second layer of Zn-tetrahedra (such as ABCD) with the same orientation as in the first layer

2nd layer of Zn-tetrahedra with a different orientation

Zn

first layer of S-centred Zn-tetrahedra all with the same orientation

S

cubic cell of a Zn atoms face-centred cubic array

in DIAMOND Zn & S are replaced by C atoms

cristobalite (cubic SiO_2)

Si-O-Si lines almost straight

B

A silicon

oxygen

C

D

oxygen

Tridymite (hexagonal SiO_2)

Si

two "cristobalite" networks without bonds between each other

oxygen
hydrogen

compare with tridymite network

Cu

Cu

oxygen

cuprite Cu_2O

H_2O-ice

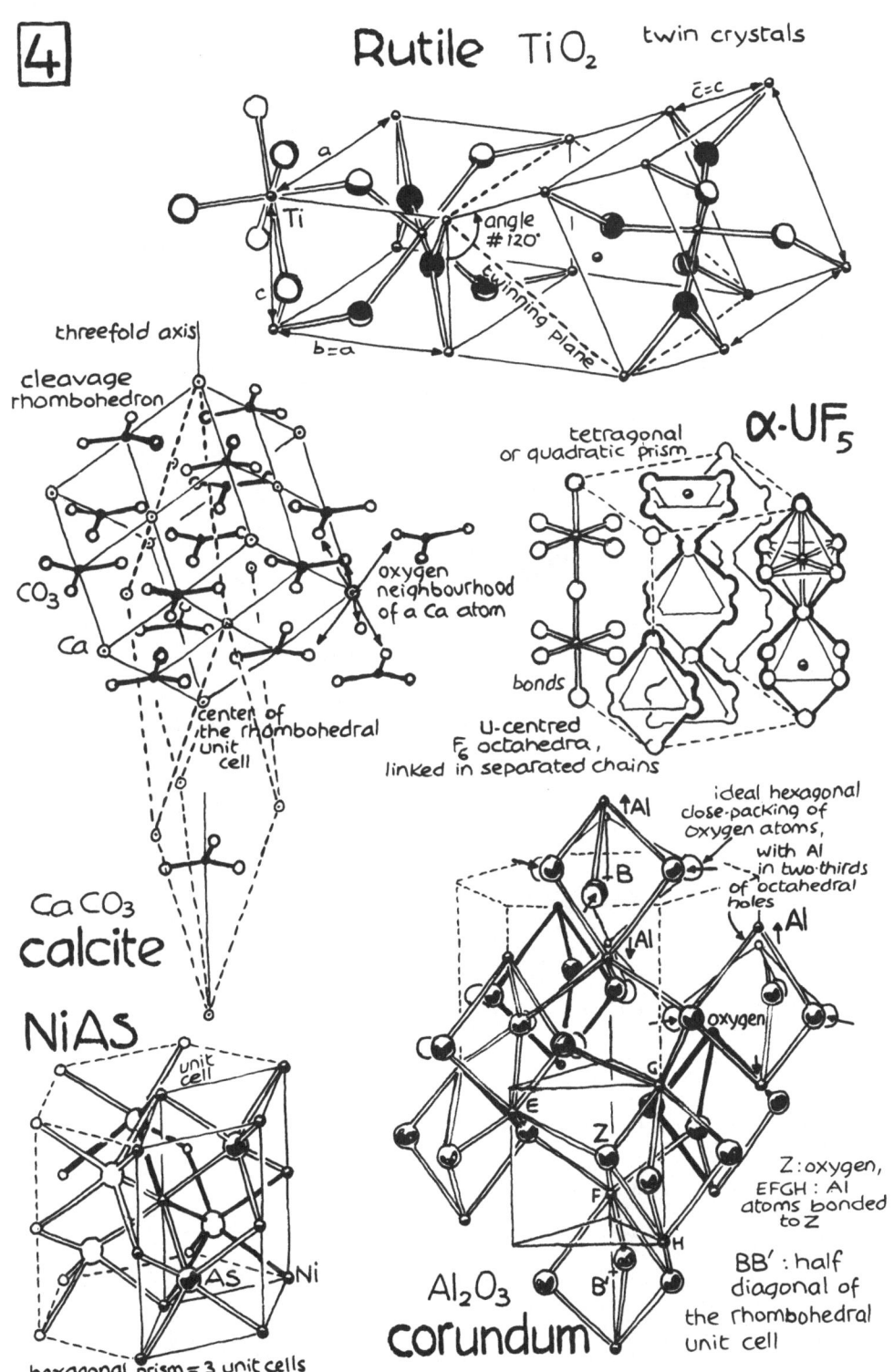

4

Rutile TiO₂ twin crystals

$\bar{c}=c$

a

Ti

angle #120°

c

twinning plane

b=a

twinning plane

threefold axis

cleavage rhombohedron

tetragonal or quadratic prism

α-UF₅

CO₃

oxygen neighbourhood of a Ca atom

Ca

center of the rhombohedral unit cell

bonds

U-centred F₆ octahedra, linked in separated chains

CaCO₃ calcite

ideal hexagonal close-packing of oxygen atoms, with Al in two-thirds of octahedral holes

↑Al

B

↑Al

Al

NiAs

oxygen

unit cell

G

E

Z

Z: oxygen, EFGH: Al atoms bonded to Z

As → Ni

F

H

BB': half diagonal of the rhombohedral unit cell

Al₂O₃ corundum

B'

hexagonal prism = 3 unit cells

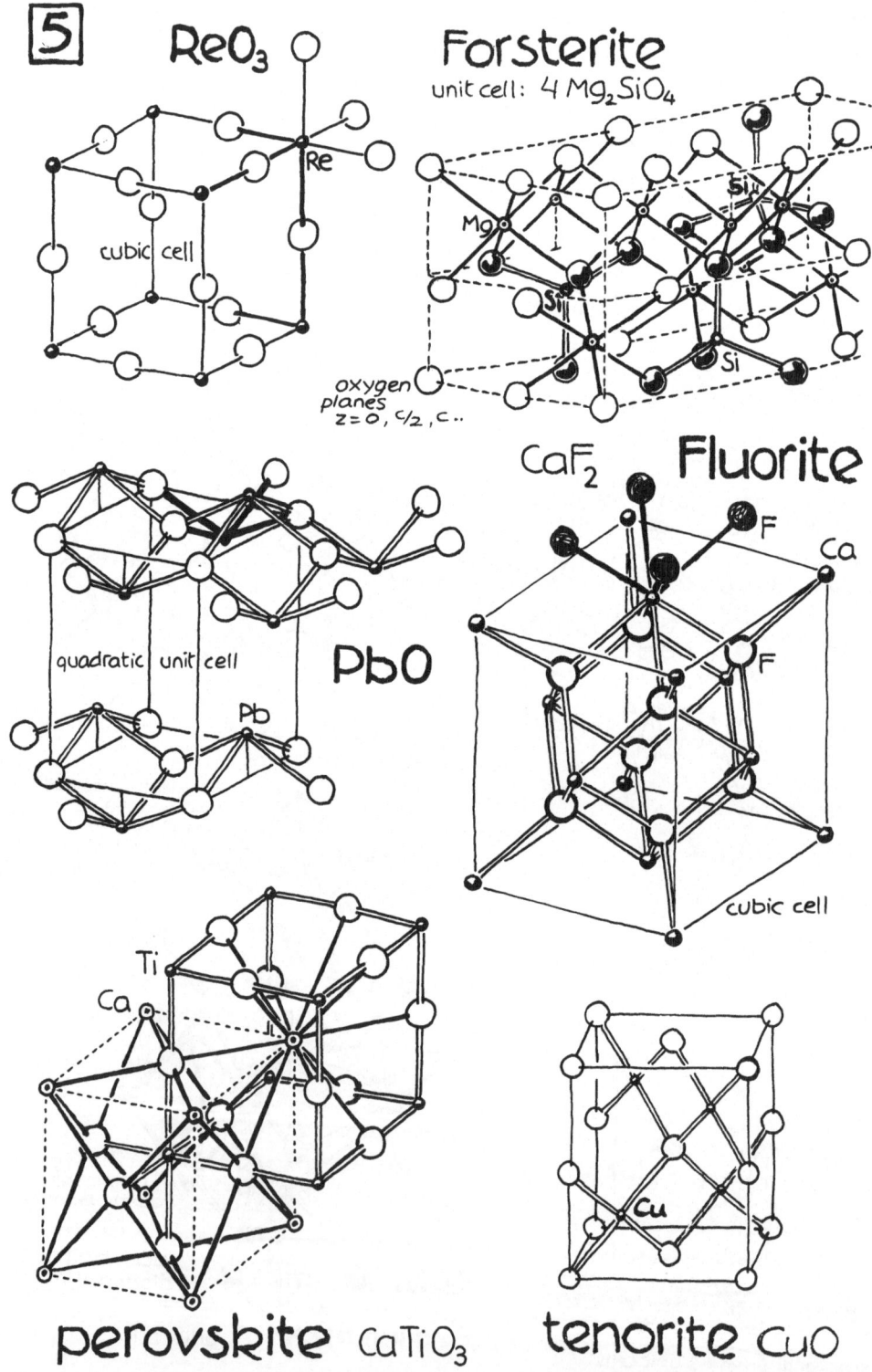

5 ReO₃ Forsterite
unit cell: 4 Mg₂SiO₄

ReO₃
cubic cell
Re

Forsterite
Mg
Si
Si
Si
oxygen planes
z=0, c/2, c..

quadratic unit cell PbO
Pb

CaF₂ Fluorite
F
Ca
F
cubic cell

perovskite CaTiO₃
Ti
Ca

tenorite CuO
Cu

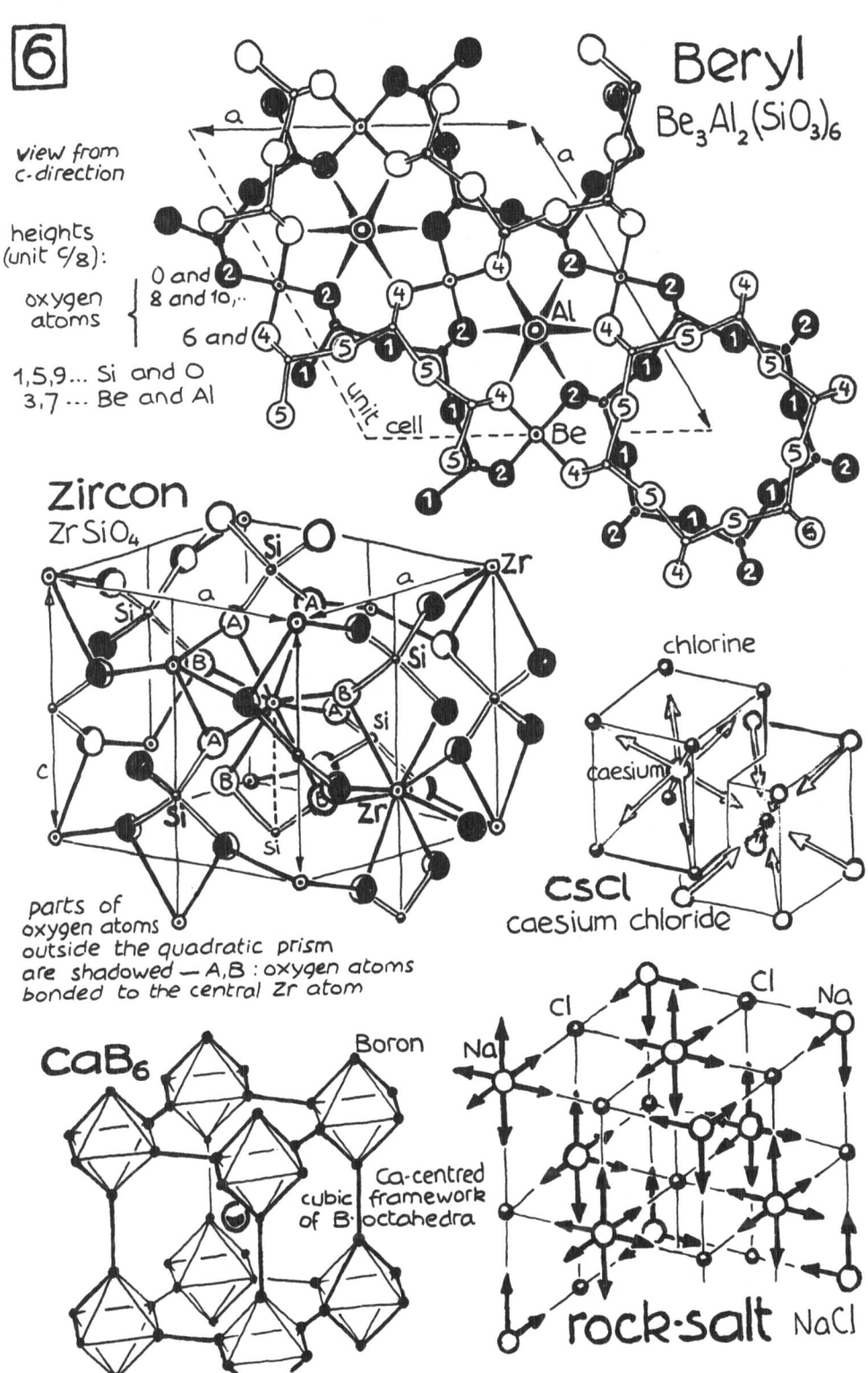

6

Beryl
$Be_3Al_2(SiO_3)_6$

view from
c-direction

heights
(unit $c/8$):

oxygen
atoms { 0 and 2
8 and 10,··
6 and

1,5,9... Si and O
3,7 ··· Be and Al

Al

Be

Zircon
$ZrSiO_4$

Si

a

Zr

A

B

Si

c

B

A

Si

Zr

Si

parts of
oxygen atoms
outside the quadratic prism
are shadowed — A,B : oxygen atoms
bonded to the central Zr atom

chlorine

caesium

CsCl
caesium chloride

CaB₆

Boron

Ca-centred
framework
of B-octahedra

cubic
framework
of B-octahedra

Cl

Na

Cl

Na

Na

rock·salt NaCl

4. Note: The Coordination Number

A remark should be made on the meaning of the expression "coordination number" used in the chemistry of finite molecules as well as in crystallography.

In chemistry, the coordination number is the number of neighboring atoms to which an atom is bound. For example, in SF_6 the coordination number of S is 6; in PCl_5, that of P is 5, etc. However, for structures in which the number of neighbors is small, the term _valence_ is used: tetravalency of carbon in CH_4, divalency of oxygen in H_2O, etc. In fact, the number of atoms linked to a given atom is equal to the number of localized molecular orbitals that can be constructed between the given atom and its immediate neighbors. As shown previously, that number results not only from the hybridization state of the atom under consideration, but also from the number of available electrons. In CH_4, NH_3, H_2O, the C, N, O atoms are sp_3--hybridized, and therefore all dispose of four free orbitals; however, because of differences in the numbers of available electrons, doubly utilized orbitals appear: one in NH_3, two in OH_2.

In molecules with more complicated hybridizations, nonbonding orbitals also play a role: in SF_6, the maximum number of bonding orbitals is actually reached, whereas in IF_5 the number of atoms linked to I is 5.

Thus the coordination number, defined as the number of bonds issuing from the atom under consideration, is not very useful to describe the structure of a molecule.

The same is true for crystals. For instance, it has been seen that in Mn_2O_3, Mn has six neighbors. Its coordination number will be assumed to be six, even though its state of hybridization (sp_3d_3f) allows formation of eight hybrids. The case of alkali halides — NaCl or CsCl — further emphasizes the difference between the notion of coordination number used in crystallography and the number of localized molecular orbitals in which the atomic orbitals of the atom under consideration participate.

The state of hybridization is certainly a more precise characteristic, provided that the number of nonbonding orbitals is specified. In this way, it is possible to know the number of nonbonding orbitals, as well as the degree of localization of the molecular orbitals, and to distinguish perfectly localized structures as in Be_2SiO_4 or partially localized ones as in NaCl.

5. Cohesive Energy in Tridimensional Macromolecular Crystals

Several cases are to be considered, according to the degree of localization that can be obtained for the molecular orbitals.

1) Crystals with localizable orbitals

The situation is the same as in a molecule of finite size. In $A_m B_n$ crystals, the total energy can be partitioned [13] in the terms associated with the two types of localized orbitals (formula 14). As in molecules, the contribution assigned to each bond depends on the rest of the structure. In crystal, owing to the periodicity of the lattice, the εs associated with the couples of bonded atoms, can, without objection, be considered as the energies of the corresponding bonds. If the distance a tends to infinity and the net charges to zero, formula (14) gives the energy of isolated atoms in the valence state that has been used to build the crystal. This valence state is, in general, an excited state. For example, in diamond, the carbon atoms are in the state (sp_3, V_4), situated at 8.14 eV above the ground state [7]. To obtain the dissociation energy from isolated neutral atoms, it is necessary to take into account the correction.

If we neglect the Madelung term, the terms ε_{ij} in crystals have the same order of magnitude as in finite molecules, built from the same basic orbitals. For example, in diamond, the localized orbitals are practically identical to those in saturated hydrocarbons (cyclohexan). It is known that the bond energies in small molecules are between 50 and 100 kcal ($\sim$ 100 kcal in C-C or O-H bonds). In a polar crystal, the Madelung term comes in addition. Consequently, the cohesive energy of a tridimensional macromolecular crystal possessing localizable orbitals is usually high. The scattering of the elements caused by an increase in temperature (melting or sublimation) is more difficult than for a molecular crystal. The melting point is always high: diamond is practically nonmelting, cristobalite melts at 1710°C, corundum at 2050°C, the silicates between 500° and 1000°C.

2) Crystals possessing nonlocalizable orbitals

The typical case is that of metals. The total energy cannot be partitioned into bond contributions because the orbitals cannot be localized. One considers the energy per atom.

In cases where the localization of the molecular orbitals leads to more or less extended delocalization cells, the difficulty is similar (e.g., in NaCℓ). It is often even more complex in mixed systems, where there are simultaneously localized and delocalized systems. We shall treat farther the case of ionic crystals.

In both molecules and crystals, the delocalization of the molecular orbitals stabilizes the system. Nevertheless, in metals, the values of the cohesive energies per atom cover a wide range: from 15 kcal in mercury to 200 kcal in tungsten. Consequently, the melting temperatures are also very different: from -39°C for mercury to > 5000°C for osmiun. The broad range of the values of the cohesive energy is due to the variety of orbitals that intervene in the lattice. Consequently, in each case, the direct calculation of the energy is necessary.

Several calculations have been performed, particularly in the case of alkaline and alkaline-earth metals, using various methods: i.e., molecular orbital method [24-25], alternant molecular orbitals method [26], localized wave functions of the Wannier type [27]. The molecular orbital method gives for the sodium [24] a value (25 kcal) very close to the experimental value (26 kcal). Nevertheless some authors [28] think that this agreement is probably coincidental since no correlation was included.

A comparative study between the classical molecular orbital method and the more improved method of alternant molecular orbitals [29], performed for lithium, shows that both procedures are in fact equivalent for interatomic distances close to the equilibrium distances in the crystal. Nevertheless, the cohesive energy obtained is equal to zero ! This result is very depressing. The cohesive energy is in fact less than 1/100 of the total energy. Therefore, it is very difficult to obtain a good value, even using sophisticated methods.

Since our purpose is not a detailed study of methods for calculating cohesive energy, a simplified semi-empirical method [30], deduced from a rigorous SCF formalism, is, for the reasons given below, largely sufficient. Owing to suitable parametrization, such a method allows a general understanding of metal structures.

Let us envision a metal possessing only one valence electron (alkalines, copper, silver, gold). All other electrons are assumed to be localized in the inner-shell orbitals. The electron charges carried by the atoms being equal to unity, the total energies of the system (taking into account the nuclear repulsion) can be practically written:

$$E = 2 \sum_i e_i - \frac{N}{4} J_{rr} + \frac{1}{2} \sum_{(rs)} \ell_{rs}^2 J_{rs} \qquad (19)$$

N = number of atoms, e_i = root of secular equation in the Hückel method (Appendix A), J_{rs} = Coulomb integral between χ_r and χ_s atomic orbitals, ℓ_{rs} = bond order between $\underline{r}$ and $\underline{s}$ atoms.

If one introduces the monocentric integrals α_r, all of which are equal to α, and the bond integrals β_{rs}, the total energy becomes:

$$E = N(\alpha - \frac{1}{4} J_{rr}) + 2 \sum_{(rs)} \ell_{rs} \beta_{rs} + \frac{1}{2} \sum_{(rs)} \ell^2_{rs} J_{rs} \qquad (20)$$

If the atoms are infinitely separated, $\ell_{rs} = 0$. Consequently, the cohesive energy is:

$$E_c = \sum_{(rs)} (2\ell_{rs} \beta_{rs} + \frac{1}{2} \ell^2_{rs} J_{rs}) \qquad (21)$$

The β and J integrals decrease rapidly when the corresponding interatomic distances increase. A good approximation consists of taking into account only nearest neighbor atoms. In abbreviated form, we shall write ℓ, β, J without subscripts. Let ν be the number of nearest neighbor atoms in the considered structure. The number of bonds (rs) is equal to $N\nu/2$. Consequently, the cohesive energy per atom is:

$$\varepsilon_c = \nu\ell\beta + \frac{\nu\ell^2}{4} J \qquad (22)$$

The empiric Hückel method should be given for the electronic bond energy: $\nu\ell\beta$. In fact, the exact expression of electronic energy is more complex. The nuclear repulsion compensates in part for the insufficiency of the Hückel method, since the second term $(\nu\ell^2 J/2)$ is weak with respect to the first $(\nu\ell\beta)$. We shall see further that the ratio $\ell J/2\beta$ is close to 0.03. Consequently, in order to obtain qualitative results, it is generally sufficient to consider only the first term $(\nu\ell\beta)$.

As proof, we shall show that it is easy to find the observed gradation for the cohesive energy in the alkaline metals. We shall assume that the bond integrals β vary according to the law generally utilized in molecules [31]:

$$\beta_{pq} = k \frac{\alpha_p + \alpha_q}{2} S_{pq} \qquad (23)$$

i.e., where: $\qquad\qquad \beta = k \alpha S \qquad\qquad\qquad (24)$

In the calculation, we shall utilize Slater's type of orbitals with suitable effective charges [7]. The distances between nearest neighbor atoms are the experimental distances [16]. The integrals $\alpha = (\mathfrak{I} + \mathfrak{K})/2$ are deduced from atomic spectroscopy [32] (see Appendix A).

Since the lattice of alkaline metals is homothetical (bcc arrangement), the bond orders are the same. The cohesive energy is proportional to the β integral. Table 2 indicates the obtained results.

The decrease of the cohesive energy explains the decrease of the melting point.

Table 2

Metal	a (Å)	S	α (eV)	β/k (eV)	Experimental cohesive energy per atom (kcal)	Melting point (°C)
Li	3.04	0.50	−3.0	−1.5	38	186
Na	3.63	0.40	−2.7	−1.1	26	98
K	4.60	0.35	−2.4	−0.8	22	64

The transition elements are particularly interesting. To a first approximation, we can admit that the s-orbitals give a doubly used molecular orbital system. In the Hückel method, the corresponding levels are symmetrically situated with respect to the energy of the s-atomic level [7]. Consequently, any contribution to the cohesive energy comes from the s orbitals. In a transition metal, the cohesion is due to molecular d orbitals. The reasoning used for the s orbitals is still valid. The cohesive energy will be maximum when half of the levels are used by the valence electrons. This conclusion is based on experiment which shows that the cohesive energy varies roughly parabolically with the number of d electrons, so that the maximum corresponds to the metals located in the middle of the series [33].

III. <u>Electron Charges and Ionicity in Macromolecular Crystals</u>

Thus far, we have considered the problem of crystal structure, particularly from a qualitative point of view. In this section, we shall study the repartition of the electron charges around the nuclei and the consequences that proceed from this repartition.

Although many calculations of electron charges have been performed for the more miscellaneous molecules, in comparison, few results have been published for the crystals. Nevertheless, despite the dearth of results, it is possible to derive interesting conclusions.

1. Methods of Calculating Electron Charges in Crystals

Of all the methods utilized to determine the total wave function in a crystal, the molecular orbital method is the most useful for obtaining the electron charges, because this method introduces atomic functions whose density is concentrated around the nuclei. The development of the total wave function into planar waves is badly fitted to such a calculation.

Systems in which, for reasons of symmetry, the value of the net charges is equal to zero (metals, diamond, graphite,...) having been eliminated, the determination of charges is conditioned by obtaining the wave function. Though the principle behind the method is the same as that for a molecule, the infinite size of a crystal creates many difficulties. The first endeavors were made using the empiric Hückel method. Coulson et al. [34], for example, studied a number of AB-like tetracoordinated crystals, using molecular orbitals localized between the A and B nuclei, built from sp_3-hybrid orbitals. The authors tried to use a more rigorous formalism for the BN crystal. The complexities that arose and the little difference obtained with the empirical calculation probably explain why these authors renounced this method. More recently, Harrison [35] studied such crystals using a semi-empirical method with a similar parametrization. Among the ab-initio-SCF studies performed in order to determine the electron charges or the values of these charges, the paper of Grimley on LiF [36] and our studies on alkali halides and on tetracoordinated crystals [13] should be mentioned.

In order to avoid the difficulties arising from the infinite size of the crystal, a classical approach consists of studying clusters that become larger and larger, but that are of finite size. By extrapolation of results obtained for the atoms situated in the center of the cluster, it is possible to estimate the charges in the infinite crystal. The quality of the result depends on the rapidity of the convergence. According to certain authors [37], about thirty atoms should be sufficient to obtain a good idea of the phenomena. In fact, the convergence depends upon the system studied [38]. This method has been used for SiO_2 [39].

Regardless of the method used, the electron charges are generally derived from the Mulliken formula (15). Even if density maps are given [40], numerical integration to obtain directly the charges located around the nuclei is rare [22].

Finally, it should be noted that calculations performed for any crystal thought to be perfectly ionic, e.g., $NaC\ell$ [41], LiF [42], $NaNO_2$ [43], ..., which produce excellent results for the energetic and mechanical properties (cohesive energy, compressibility), constitute an indirect proof of the existence of net charges practically equal to integer numbers, carried by the atoms or groups of atoms (in $NaNO_2$). These calculations can be considered as a posteriori methods of determination of charges.

2. General Results

There is a greater variation in the values obtained in crystals for the electron charges than in those obtained in molecules. All the intermediate values become: from net charges equal to zero, up to ionic charges. To discuss the results, it is practical to classify them according to the structural types discussed above.

1) Crystals possessing localizable orbitals

Table 3 contains a list of results obtained from crystals possessing localizable orbitals built from couples of sp_3-hybrids, pointing in two's against each other. We give three sets of values, obtained respectively from an empirical method, a semi-empirical method, and a SCF calculation.

In quartz, the values of Si net charge obtained for clusters of variable size [39] are located between +3.4 and 2.8 .

Table 3

Crystal	Net Charges		
	Coulson [34]	Harrison [35]	Julg [13]
BN	0.43	0.71	0.44
BeO	0.56	0.70	0.71
AℓP	0.46	0.87	1.16
PB	-	-	1.14
SiC	0.28	1.41	1.57
BeS	0.40	-	0.22
AℓN	0.56	1.29	1.45
GaAs	0.46	0.87	-

2) Crystals possessing partially delocalized orbitals

In this category, we only have results for alkali halides. After Grimley [36], in LiF, the net charges are very close to +1 and -1. In Table 4, we indicate after [22] the increasing electron charge carried by the metal in alkali halides. These values were obtained by numerical integration of the electron density contained inside a sphere radius equal to the classical ionic radius.

Table 4

Crystal	Increase of electron charge of the cation
LiF	0.012
NaF	0.018
K F	0.017
LiCℓ	0.012
NaCℓ	0.019
KCℓ	0.013

The net charges are very close to 1. The result indicates that the a priori delocalized orbitals are, in fact, practically reduced to the atomic orbitals of the halogen atoms.

In addition, the cellular model indicates that the electron densities around the fluorine or chlorine atoms possess cubic symmetry [22]. This result is in agreement with the density maps obtained by diffraction methods [44] and is consistent with the existence of a signal in quadrupolar resonance [7].

Other methods (dynamic lattice calculations [45] or dielectric theories [46]) yield slightly higher net charges:

LiF: 0.8 - 0.9   LiCℓ: 0.8
NaF: 0.8 - 0.9   NaCℓ: 0.7 - 0.8
K F: 0.9   K Cℓ: 0.8

(Lundquist [47] indicates 0.9 for NaCℓ).

In return, the photoelectron spectra, taking into account the Madelung term in crystals, show that these same net charges are very close to unity [48]. The value of ionization energy of 1s orbitals of fluorine atoms is practically the same in CsF and KF, in which the net charge would not be expected to deviate significantly from +1, as in LiF. On the other hand, the ratio Q(KBr)/Q(KCℓ) is equal to 0.98 [49]. Consequently, we think that the net charges in alkali fluoride and chloride are very close to unity. We shall return to this point later.

3) Metals

By reason of symmetry, the net charges are equal to zero. The molecular orbital method has been applied to alkaline metals [50]. For net charges, a value equal to zero has evidently been obtained. More interesting is the case of metals possessing many valence electrons. The total electron charge is, of course, equal to the number of valence electrons, but the repartition of the electrons in the orbitals is nonuniform. Calculation of the atomic orbital populations in transition metals shows that the electron cloud is not spherical and that it possesses the symmetry of the lattice. Table 5 indicates the populations of the d orbitals in nickel (cfc), iron (bcc), and cobalt (hcp) [51].

Table 5. Population of the d orbitals

Metal	Orbitals: xy	yz	zx	$x^2 - y^2$	z^2
Ni	1.74	1.74	1.74	1.89	1.89
Fe	1.32	1.32	1.32	1.52	1.52
Co	1.59	1.63	1.63	1.59	1.56

Nevertheless, the asphericity is weak. We shall see farther the effect of faces.

3. Definition of Bond-Ionicity in a Crystal

In a crystal with localized orbitals, the presence of nonvanishing net charges around the nuclei can be interpreted as arising from the dissymmetry of partial density distributions $|\varphi_i|^2$ corresponding to various localized orbitals. To specify this dissymmetry, Coulson et al. [34] proposed introduction of the _ionicity of the bond_:

$$\alpha = \frac{|q_a - q_b|}{q_a + q_b} \tag{25}$$

where q_a and q_b are the electron populations, respectively, carried by the orbitals χ_a and χ_b used to build the corresponding molecular orbital φ_{ab}. If the repartition is symmetric, as in diamond, $q_a = q_b$ and $\alpha = 0$, and the crystal is not ionic. If $q_a = 0$, e.g., $q_b = 2$, $\alpha = 1$, and the crystal is perfectly ionic.

The _bond-ionicity_ so defined is not identical with _atom-ionicity_. For example, in the molecule BeO, if the bond is symmetric ($\alpha = 0$), the net charges carried by the atoms are different from zero: $Q_{Be} = +1$ and $Q_O = -1$.

If the atoms A and B are linked by a simple bond, and if all the bonds are identical, $q_A + q_B = 2$. Let us assume the atom B is more electronegative than A; then the electron charge q_B is greater than q_A. Consequently Eq. (25) becomes:

$$\alpha = 1 - q_A \tag{26}$$

More generally, we shall define the average ionicity [52] of a crystal $A_m B_n$ by:

$$\alpha = \frac{|q_A/\nu_A - q_B/\nu_B|}{q_A/\nu_A + q_B/\nu_B} = 1 - \frac{q_A}{\nu_A} \tag{27}$$

where q_A and q_B are the total electron charges carried by all the atomic orbitals used to build the localized orbitals; ν_A and ν_B are the number of orbitals carried by the atoms A and B. One verifies that if the repartition is symmetric:

$$q_A/\nu_A = q_B/\nu_B \quad , \quad \alpha = 0 \quad ,$$

and if $q_A = 0$, e.g., $\alpha = 1$.

In the case of NaCℓ , where it is not possible to build localized molecular orbitals, the previous formula can still be utilized. Now, ν_A (or ν_B) is the number of atoms B (or A) which carry atomic hybridized orbitals pointing to A (or B).

In a more general way, for all the $A_m B_n$ crystals in which A is a metal and B is a polynuclear anion having a strong covalent structure (carbonate, nitrate, sulfate, silicate, chlorate ...), relation (27) is valid. Q_A^{io} being the net charge of an atom A in a crystal assumed to be perfectly ionic, we have:

$$Q_A = Q_A^{io} - q_A \tag{28}$$

Consequently:

$$\alpha = \frac{Q_A + \nu_A - Q_A^{io}}{\nu_A} \tag{29}$$

For example, in fluorite CaF_2: $(Q_{Ca}+6)/8$, in phenacite Be_2SiO_4: $(Q_{Be}+2)/4$, in forsterite Mg_2SiO_4: $(Q_{Mg}+4)/6$, etc.

In more complex crystals, such as beryl, one defines local ionicities. For the orbitals between Be and O: $(Q_{Be}+2)/4$, for the ones between Aℓ and O: $(Q_{Aℓ}+3)/6$.

In Table 6 are listed the values of the ionicities corresponding to the charges indicated in Table 3. In alkali halides, α is practically equal to 1.

Table 6

Crystal	Bond-ionicity		
	Coulson	Harrison	Julg
Diamond	0	0	0
BN	0.36	0.43	0.36
BeO	0.64	0.68	0.68
AℓP	0.37	0.47	0.54
PB	0.33	—	0.53
SiC	0.06	0.35	0.39
BeS	0.60	—	0.56
AℓN	0.39	0.57	0.61
Si	0	0	0
GaAs	0.46	0.87	—

The values obtained by the SCF method or the semi-empirical method of Harrison, do not differ significantly. However, these values differ from those previously proposed by Coulson et al.

4. Application of the Ionicity Notion to the Determination of the Hardness of a Crystal [52]

The hardness of a crystal (Appendix D) is related to its structure. Indeed the hardness depends on the facility to displace or remove the superficial elements of the sample.

A molecular crystal (or one composed of macromolecular layers) is soft because there are no molecular orbitals joining various entities of the crystal. In a tridimensional macromolecular crystal, the existence of a framework, which sustains the structure, confers a greater hardness to the crystal.

In diamond, for example, the displacement of a carbon atom is impossible without breaking four molecular orbitals. This phenomenon results from the important directional character of hybrids. The necessary energy is great. Diamond is hard (10 on the Mohs scale). In contrast, in sodium chloride the molecular orbitals that assure the cohesion of the lattice are practically reduced to s and p orbitals of chlorine atoms. The corresponding density is practically spherical. The displacement of a sodium atom is facilitated by the isotropy. NaCl is soft: H = 2.5.

However, it should not be concluded that the purely covalent crystals are necessarily harder than crystals that are partially ionic in character. In fact, the facility with which atoms can be displaced also depends on the facility with which the bonding molecular orbital can be deformed by external perturbation. Although it does not exactly correspond to the same phenomenon, the facility of mechanical deformation of the framework can be thought to be related to the electrical polarizability of bonds, because in quantum mechanics the operators corresponding to interactions between two systems only introduce electrostatic potentials. It is now known that bonds between atoms belonging to the same row have comparable polarizabilities, and that the polarizability increases considerably when these atoms are replaced by one or two atoms belonging to the next row [7]:

C-H	1.7	C-C	1.2	C-S	4.7
O-H	1.9	C-N	1.7	C-Cℓ	7.3
N-H	1.8	C-O	2.1	C-Br	12.0
S-H	4.7	C-F	2.6	C-I	17.3

Consequently the hardness decreases. This explains why silicon is much less hard than diamond despite its covalent structure ($\alpha = 0$): $H_{Si} = 7$.

The hardness appears to be related to two factors: the ionicity of the bonds and the polarizability of the bonds. We shall suppose a general relation of the type:

$$H = (\text{Function of } \alpha) \times (\text{Function of polarizability}) \qquad (30)$$

Since the polarizability depends essentially on the period to which the atoms belong, one can attempt to determine, for the expression of the hardness of a crystal $A_m B_n$ with localizable orbitals, a relation of the type:

$$H_{pq} = K_{pq} \, f(\alpha) \qquad (31)$$

where p and q are the numbers of the rows to which atoms A and B belong, and K_{pq} is a constant ($p = 2$: Li, Be, ... ; $p = 3$: Na, Mg, ... etc).

By combining the experimental and theoretical values presently available, a simple formula is obtained:

$$H_{pq} = K_{pq}(1 - \tfrac{2}{3} \alpha^4) \qquad (32)$$

The values of K_{pq} ($= K_{qp}$) are indicated in Table 7. The increase of polarizability with increasing values of p and q explains the gradations obtained.

Table 7

Values of p and q used in formula (32)

q \ p	2	3	4	5
2	10.0	9.5	9.0	8.6
3	—	7.0	6.7	
4	—	—	6.0	

In Table 8, the values for hardness calculated from formula (26) are compared with experimental values for various types of crystals. For gallium arsenide, we used experimental [53] and theoretical [34-35] results that indicate a net charge for the gallium atom of about 0.3-0.5.

The experimental values of the hardness are taken from references [54-55], after smoothering [52] (Appendix D). The value of hardness of GaAs was determined by myself.

On the whole, the agreement is satisfactory. Formula (32) gives hardness with a maximum error of 0.1.

Assuming an error of 0.1 for $\underline{H}$, the various values of the constant K_{pq} can be given by the unique formula:

$$K_{pq} = 5 + \frac{6}{p-1} - \frac{q}{p} \qquad (p \leqslant q) \qquad (33)$$

Consequently the hardness of a $A_m B_n$ crystal is given by:

$$H_{pq} = (5 + \frac{6}{p-1} - \frac{q}{p})(1 - \frac{2}{3} \alpha^4) \qquad (p \leqslant q) \qquad (34)$$

Table 8

Crystal	Net charge	Ionicity	Calculated hardness	Observed hardness
Diamond	0	0	10.0	10.0
BN	0.44	0.36	9.8	9.5-10
BeO	0.71	0.68	8.6	8.5-9
LiF	∿1	∿1	3.3	3.3
SiC	1.57	0.39	9.2	9.2
AℓN	1.45	0.61	8.6	8.5-9 .
NaF	∿1	∿1	3.2	3.2
LiCℓ	∿1	∿1	3.2	3.2
Si	0	0	7.0	7.0
NaCℓ	≲1	≲1	2.5	2.5
KCℓ	∿1	∿1	2.2	> 2
Ge	0	0	6.0	∿6
GaAs	0.3-0.5	≲1	5.9	≲6

In crystals in which $\underline{B}$ is a group of atoms (carbonate, silicate, ...) that can be considered sufficiently rigid due to their

structure, the hardness will depend only on local ionicity around
the metal. Formulas (32) and (34) would therefore be applicable to
them. Unfortunately, the values of charges in such crystals are not
available. Only the $NaNO_2$ case [43] seems to show that the net charges
are practically ±1. On other hand, one can assume that in the alkali
or alkali-earth carbonates and sulfates, the net charges also are
integers, because of the large differences in electronegativity
between oxygen and these metals. Assuming this hypothesis, the
hardness of these crystals is $H_{pq}/3$. Table 9 is a comparison of the
values of $H_{pq}/3$ with the observed experimental values.

Table 9

Crystal		Hardness calculated with $\alpha = 1$	Observed hardness
Thenardite	Na_2SO_4	3.0	3
Magnesite	$Mg\ CO_3$	3.3	3.5-4
Calcite	$Ca\ CO_3$	3.0	3.2
Anhydrite	$Ca\ SO_4$	3.0	3-3.5
Celestite	$Sr\ SO_4$	2.9	3-3.5
Baritite	$Ba\ SO_4$	2.7	2.5-3.5
—	$K\ C\ell O_3$	3.0	$\leqslant 3$

The calculated hardnesses are somewhat less than the observed
values. The results are consistent because the values obtained with
$\alpha = 1$ represent the lower boundaries for hardness. In addition, the
small differences between calculated and observed values constitute
a sound argument in favor of formula (34). It should be noted that
the values indicated for celestite and baritite were obtained by
assuming that formula (28) is applicable up to a value q = 6.
Moreover, the X-ray absorption [56] gives for the zirconium atom in
zircon ($ZrSiO_4$) a net charge of 1.6. According to formula (34), the
hardness would be about 7.2. The experimental value is located
between 7.0 and 7.5.

Nevertheless, it should be noted that, according to the above
reasoning, an _average_ value of hardness for a crystal has been
assumed without taking into account the possible effects of
anisotropy. In a crystal, the arrangement of the constituent
elements is different in the various cleavage planes; thus, the

hardness will depend on the particular face and direction under consideration. In general, the differences arising from anisotropy are small (e.g., in calcite). Disthene, in which hardness varies between 4-5 and 6-7 for the face (100) depending on the direction [54], is an exceptional case. This phenomenon probably occurs because of the presence of chains of tetrahedron SiO_4 on the surface that are broken more easily in the perpendicular direction than in longitudinal direction.

5. Application to the Determination of Electron Charges

A comparison between the hardness calculated using formula (34) and the observed values seems sufficiently satisfactory to hope that the charges obtained from observed hardness, if not particularly accurate, are of a reasonable order of magnitude. Indeed several factors reduce the accuracy. Independently of the fact that relation (34) is nothing but an approximation, a great uncertainty arises due to the term α^4. For crystals with weak ionicity it is difficult to obtain the value of the electronic charges with sufficient accuracy.

Table 10 gives net charges obtained for some minerals.

In corundum, a difficulty occurs owing to the fact that we have two kinds of Al-O bonds, three of length 1.97 $\overset{o}{A}$, and three of length 1.86 $\overset{o}{A}$. Likewise in rutile: Two Ti-O bonds of length 1.98 $\overset{o}{A}$ and four Ti-O bonds of length 1.95 $\overset{o}{A}$. The differences are weak, we shall assume that all the bonds are equivalent.

Table 10

Crystal		Hardness	Estimated charge of the metal	Ionicity
Periclase	MgO	5.1	1.5	0.91
Calcium oxide	CaO	4.5	1.6	0.93
Strontium oxide	SrO	3.5	1.8	0.97
Barium oxide	BaO	∿3.3	1.8	0.97
Zincite	ZnO	5.9	1.4	0.85
Tenorite	CuO	∿5	1.6	0.90
Sphalerite	ZnS	4.0	1.5	0.88
Galena	PbS	3.0	1.6	0.93
Corundum	Al_2O_3	8.9	∿0.3	0.53
Quartz	SiO_2	7.0	3.2	0.79
Rutile	TiO_2	6.5	∿2.8	0.80
Sellaite	MgF_2	∿5.3	1.4	0.90
Fluorite	CaF_2	3.7	1.8	0.88
Nantokite	CuCl	2.8-3	∿0.8	∿0.96
Marshite	CuI	2.8-3	∿0.8	∿0.96
Calcite	$CaCO_3$	3.3	1.9	0.99
Aragonite	$CaCO_3$	3.6	1.8	0.97
Smithsonite	$ZnCO_3$	5.5	1.2	0.87
Cerusite	$PbCO_3$	3.5	1.7	0.97
Strontianite	$SrCO_3$	3.8	1.6	0.96
Whiterite	$BaCO_3$	3.6	1.6	0.95
Anglesite	$PbSO_4$	3-3.3	1.9	0.98
Phenacite	Be_2SiO_4	7.5-8	∿1	∿0.72
Forsterite	Mg_2SiO_4	∿6.5	∿1	0.80
Willemite	Zn_2SiO_4	5.5	1.5	0.72
Perowskite	$CaTiO_3$	5.7	1.5	0.90
Geikielite	$MgTiO_3$	5-6	∿1.4	∿0.90
Crocoite	$PbCrO_4$	3.0-3.3	1.8	0.98
Scheelite	$CaWO_4$	∿5	∿1.2	∿0.90
Zircon	$ZrSiO_4$	7-7.5	∿1.6	∿0.7

6. Discussion of the Results

The theoretical results obtained for some crystals (Table 6) and the results obtained from the empirical formula (34) (Tables 8-10), permits a sufficiently extensive discussion.

First, in $A_m B_n$ crystals with localizable orbitals, all degrees of ionicity are possible, from 0 in diamond or silicon, to about 0.9 in CaF_2, MgF_2, ZnO, ZnS.

Two factors seem to play a role in the displacement of electron charge in localized orbitals φ_{ij} [13]: (1) the difference in electronegativity, Δx, between the atoms, and (2) their polarizability. The greater the difference in electronegativity, the greater is the displacement of the charge toward the more electronegative atom. In addition, the polarizability of an atom increases with its size. Consequently, the atom holds the electrons less strongly. In a given row of the periodic table, the atom polarizabilities are of a comparable order of magnitude, whereas they increase notably from one row to the next [57]. This explains why BeO (where $\Delta x = 1.9$) is more ionic than BN (where $\Delta x = 1.0$). In ZnS, the electronegativity difference ($\Delta x = 0.9$) is much less than in ZnO ($\Delta x = 1.8$), but the sulfur atom is more polarizable than the oxygen. The two factors oppose each other, but ZnS and ZnO possess comparable ionicities. The same is true in AℓN ($\Delta x = 1.4$) and AℓP ($\Delta x = 0.6$).

It is interesting to note that in oxides and sulphides, when the ionicity increases, the structure with localized orbitals becomes a structure with cells centered on the nonmetal. The critical point seems to be situated a little above $\alpha = 0.9$.

In comparison with the charge obtained for the aluminum atom in corundum, the charge of silicium in quartz seems to be high, considering the hardness of this mineral. However, this value is in good agreement with those obtained for clusters of increasing size: 2.8 to 3.4 [39]. The high value of the ionicity is probably related to the piezoelectronic properties of quartz.

In crystals where $\underline{B}$ is a complex group of atoms, we see that in the sulfates and probably also in the nitrates, the ionicity is not as high. By the way, note the weak difference between the structures of calcite and aragonite. For a given metal (e.g., Ca) the charge seems less in the aragonite structure.

In silicates, titanates, and tungstates, despite important differences in electronegativity between the oxygen and the metal, the ionicity is lower (Table 11). A clean break seems to exist between the two categories of salts.

Table 11

Ionicity in some oxisalts

Metal	Δx O-M	Ionicity						
		CO_3	SO_4	$C\ell O_3$	CrO_4	SiO_4	TiO_3	WO_4
Na	2.5	−	1.0	−	−	−	−	−
K	2.6	−	−	1.0	−	−	−	−
Be	1.9	−	−	−	−	0.72	−	−
Mg	2.1	0.99	0.97	−	−	0.80	0.90	−
Ca	2.4	0.99-0.97	0.98	−	−	−	0.90	0.90
Sr	2.5	0.96	0.98	−	−	−	−	−
Ba	2.6	0.95	0.99	−	−	−	−	−
Zn	1.8	0.87	−	−	−	0.72	−	−
Pb	1.1	0.97	0.98	−	0.98	−	−	−

In copper halides, CuCℓ and CuI, the ionicity is much less than in the alkali halides. The net charges should be only 0.7 or 0.8. These values are higher than those proposed by Coulson [34] ($\sim$0.4) or by Harrison [35] ($\sim$0.1).

Nevertheless, the lack of hardness and the rock-salt structure should suggest an ionic lattice.

The semi-empirical determination from dynamic lattice calculations leads in MgO, CaO, SrO, ZnS, CaF_2 and TiO_2, to net charges equal to, respectively: 1.8, 1.6, 1.2, 1.0, 1.5, and $\sim$3. In CuCℓ and CuI, the values obtained are practically equal to 1. Though we have seen that in the case of the alkali halides, the method gives only approximate values (in general, rather weak), the agreement between our values and those obtained by this method is acceptable.

Finally, we shall return to the case of beryl, which contains two metal atoms. Formula (34) is not directly applicable. Moreover, the obtained results permit the conclusion that the ionicity of Aℓ-O bonds is greater than that of Be-O bonds. In both cases, the electronegativity differences are practically the same (1.8 and 1.9), but the aluminum atom is more polarizable. Consequently, the hardness is conditioned by the ionicity of Be-O bonds, which are the strongest. If we assume that the environment of Be atoms is practically the same

as in phenacite, beryl should have the same hardness. This prediction
is conform to experimental value.

All the results show the validity of the theory.

7. The Ionic Approximation

Traditionally, the classification of crystals includes an
important category: the ionic crystals. These crystals are composed
of ions, and the cohesion arises from the balance between the
attractive coulombian forces and short-distance repulsive interactions,
which prevents the collapse of the crystal [58]. This category includes
the alkali halides, all crystals possessing a similar structure (MgO,
CaO), the oxy-salts (carbonates, nitrates, sulfates, silicates).
Sometimes even corundum is included in this category!

In fact, we have seen that, in certain salts (e.g., alkali
halides) the net charges are very close to those in the isolated ions.
In contrast, in the other cases, appreciable differences appear:
1.9 in calcite, 1.8 in fluorite. The classical ionic model does not
consider these charge transfers between ions. Consequently, it cannot
give a correct description of crystals, except when the transfers are
very weak, e.g., alkali halides. In that case, it makes no difference
whether ions ($C\ell^-$ or Na^+) or completely localized molecular orbitals
around the nuclei are used to build the total electron wavefunction.

We showed that, even in alkali halides, an electron transfer
always exists. Consequently, the ionic model is not entirely
satisfactory. It serves only as a special case of the general models
proposed previously.

In part B, we covered molecular crystals, lattices composed of
ions (e.g., phosphorus pentahalides). From a conceptual point of view,
it is important to emphasize clearly that, despite the apparent
similarity with practically ionic crystals ($NaC\ell$), a fundamental
difference exists between the two types. In $PC\ell_4^+$, no electron
transfer is possible from $PC\ell_6^-$ toward $PC\ell_4^+$, because all the basic
orbitals of the chlorines are doubly used in these ions. In contrast,
in $NaC\ell$, an electron transfer is possible, even in the elementary
description, which uses a minimal basic set of orbitals. Though the
transfer is very weak or even negligible, such crystals cannot be
included in the same category as those in which the transfer is
impossible.

8. Deformation of Ions in the Ionic Model

If we agree to neglect charge transfers when they remain weak
(e.g., $\leqslant 0.1$), many crystals can be considered as being built up from
ions. This is true for alkali halides, alkali and alkali-earth
carbonates, nitrates, and sulfates. Silicates are excluded from this
category, as are the magnesium and calcium oxides.

The exact electron charge carried by various entities in the
so-called ionic crystals notwithstanding, we can ask whether or not,
as in molecular crystals where molecules remain minimally perturbed,
ions possess properties, if not identical, at least similar to those
expressed when they are isolated.

Table 12 contains a list of the results obtained for the charges
of the NO_2^- ion in sodium nitrite crystal [43], and Table 13 includes
the results corresponding to calcite, magnesite, lithium, and sodium
nitrates [59].

Table 12

Net charges in nitrite ion

NO_2^-	Q_N	Q_O
In vacuo	0.158	-0.579
In crystal	0.191	-0.596

Table 13

Characteristics of carbonate and nitrate ions in salts

	$CO_2^=$	$CaCO_3$	$MgCO_3$	NO_3^-	$NaNO_3$	$LiNO_3$
d_o (Å)	1.327	1.332	1.334	1.316	1.317	1.318
Q_O	-0.718	-0.728	-0.730	-0.384	-0.390	-0.390
$\tilde{\nu}_{A_1}$ (cm^{-1})	1105	1064	1064	1075	1071	1069
$\tilde{\nu}_{A_2}$ (cm^{-1})	836	863	864	683	694	695
$10^{-5}k$ (cgs)	11.7	10.8	10.8	11.1	11.0	10.9

(d_o = equilibrium distance CO or NO, $\tilde{\nu}$ = wavenumbers of first
vibrational transition, k = force constant).

The values listed in Table 12 were obtained assuming that the geometry of the NO_2^- ion is the same as that in the isolated NO_2^-. The crystal field was simulated by a suitable set of point charges +1 and -1. The use of a more sophisticated description for the crystal field, which replaces the neighboring ions by three point charges or by a continuous distribution, gives practically the same results.

The values given in Table 13 were obtained by a SCF-ab initio method (STO-6G), with a crystal field simulated by 728 point charges. The geometry has not been assigned.

The oxygen net charges increase. This phenomenon is due to the presence of positive ions in their neighborhood, which make the more important contribution to the crystal field. The equilibrium distance d_o increases and the force constants decrease. The effects are greater in carbonates than in nitrates, because the field is created by double charges.

Recent ab initio calculations on the WO_6^{6-}ion [60], in which the surrounding lattice is simulated by six or eight positive point charges, indicates, of course, a modification of electron charges. However, the number of point charges introduced is too small to provide a quantitative conclusion.

These results show that significant modifications appear, but that they can be considered as perturbations caused by the crystal field. They only weakly alter the properties of ions. In calcite or magnesite, there is a transfer from carbonate ions toward the metal (Table 10), which modifies the number of electrons for the carbonate system. To estimate the corresponding effect on the geometry, we shall use the molecular orbitals corresponding to the isolated ions [61]. The loss of 0.1 electron, which affects the two highest molecular orbitals, does not modify the charges of the carbon atom. Consequently, the negative charge of each oxygen atom decreases by 0.1/3. The bond order of the C-O bonds is not affected. However the C-O distance increases by about 0.01 Å. This effect is due to the modification of the net charges [62]. On the other hand, the crystal field decreases (multiplied by a factor 1.9/2 = 0.95). The increasing effect on the geometry is weaker. Consequently, the two effects oppose one another so that one can hope for a partial balance.

In silicates (e.g., Be_2SiO_4), where the ionicity is lower than unity, one must expect sensible modifications in SiO_4 groups with respect to the isolated SiO_4^{4-} ions. Unfortunately, simple perturbation calculations such as those performed for calcite, cannot be applied. The values given in the literature do not allow a decision as to whether the indicated modifications are due to the inaccuracy of the determinations or if they correspond to real differences in structure (as in the case of $ZrSiO_4$).

IV. The Crystal Arrangement

It is possible to predict the geometry of a molecule. Can it also be predicted for a crystal? The large number of parameters (the coordinates of nuclei) prevents an ab initio investigation of the positions of atoms in a crystal. Naturally, the problem is simplified if one assigns a general pattern to the structure. The geometry depends on only one parameter. The minimization of the energy with respect to this parameter gives the geometry of the lattice. Such calculations have been performed for alkali-halides [41-42], diamond [63], and metals [29-26]. The quality of the values obtained depends heavily on that of the wave function. In contrast, the charges seem less sensitive to this factor.

In fact, such calculations do not give a direct answer to the problem of the general prediction of crystal arrangement. This crucial problem has been solved only in specific cases. The examples below show that, without performing more sophisticated (and consequently more extensive) theoretical calculations, it is also possible, using elementary calculations or simple qualitative considerations derived from the hybridization and the localized orbital concepts, to obtain interesting results in many cases.

1. Preliminary Remarks

Before examining the question of crystal arrangement, it is necessary to know whether the substance has a molecular or tridimensional macromolecular lattice. According to what we saw, one could assume that a compound that exists in the gaseous or liquid state as a well-defined molecule, is inclined to give molecular crystals: e.g., organic compounds (hydrocarbons), metalloids (I_2, Cl_2, P_4,...), small inorganic compounds (H_2O, NH_3, CO_2 ...). In contrast, one could assume that in compounds that do not exist as small molecules (salts, oxides), the network is a macromolecular lattice. In fact, the problem is much more complex. For example, NaF and NaCl in the vapor state are molecules, but yield a tridimensional crystal. Likewise, one can ask why solid CO_2 is a molecular crystal when SiO_2 is a tridimensional lattice. One is not allowed to invoke the value of the molecular weight for SiO_2, since for example, P_4O_6 and P_4O_{10} give molecular crystals like the nitrogen oxides, which are lighter than they are. Only a calculation of both types of structures

can provide an answer to the question. The problem, which is simple
in principle, is in fact difficult to solve because the geometry of
the hypothetical lattices is not known. We shall see that in the case
of certain metals, it is easy to obtain an answer.

2. Inadequacy of the Classical Criterion Derived from the Ionic Radii Ratio

From Pauling's works, the size of the ions that are assumed to
make up the crystal is considered the most important factor in the
choice of the crystal arrangement [1]. The classical theory considers
that most crystals are composed of ions packed one upon the other.
The stability of the crystal in a such model results from the balance
between the electrostatic forces and the short-distance repulsion
forces between the ions which are assumed to be impenetrable [58]. In
the case of crystals composed of monoatomic ions, knowledge of the
interatomic distances permits the assignment of a radius to various
ions, which are assumed to be spherical: the ionic radius. Several
sets of values have been proposed: those of Pauling [1], Goldsmith [64],
and Ladd [65]. In fact, there is very little difference among the values,
and they lead to homogeneous results (see Table 14). A complete list
is given in reference 17.

The resultant of electrostatic forces is an attractive force.
The energy appears as -A/a where A is the Madelung constant of the
lattice and a the distance between two nearest neighbors. In the
absence of the short-distance repulsion, the crystal should collapse
on itself. Consequently, the more stable arrangement will be the most
compact, taking into account the sizes of ions assumed to be
impenetrable, rigid spheres. Under these conditions, one understands
the leading part that should be played by the ionic radii. For example,
in the cfc arrangement (NaCℓ), the ratio $\rho = r_A/r_B$ must be located
between 0.414 and 0.732; in the bcc arrangement: $0.732 < \rho < 1$; in the
fluorite structure: $0.414 < \rho < 0.732$; in the zinc-blende structure:
$0.224 < \rho < 0.414$.

In fact, it has long been that these conditions are not always
respected [16]. The case of alkali halides is typical. Table 14 gives
the various ratios for ρ. According to the values of ρ, only those
crystals corresponding to the underlined values would crystallize in
the cfc system, whereas those corresponding to values higher than
0.732 (indicated by a dotted line, ---) would have the bcc structure.

In fact, only CsCℓ, CsBr, and CsI, under normal conditions, possess
the bcc structure. Cesium chloride crystallizes with the rock-salt
structure (cfc) at temperatures above 445°C. This indicates that the
bcc and cfc structures have similar energies. One should note in
Table 14 the too-weak values obtained for LiCℓ, LiBr, and LiI, which
would crystallize in the zinc-blende or wurtzite structures. Finally,
let us note that the distances between nearest neighbors in the
arrangements bcc and cfc, observed in halides possessing both
structures, are practically the same [66]. AgF, in spite of the value
$\rho \sim 0.9$, crystallizes in the cfc structure.

Table 14

Values of ratios of ionic radii

	Li	Na	K	Rb	Cs
F	0.51-0.44	0.74-0.70	1.00-0.98	1.11-1.09	1.26-1.24
Cℓ	0.38-0.33	0.54-0.52	0.73-0.73	0.82-0.84	0.92-0.93
Br	0.35-0.31	0.50-0.49	0.68-0.68	0.76-0.76	0.85-0.87
I	0.31-0.28	0.45-0.44	0.61-0.62	0.68-0.69	0.76-0.78

(The first value corresponds to Goldsmith's radii, the second
to Pauling's radii)

Similar difficulties appear in oxides or fluorides (Table 15).

Table 15

Ratio of ionic radius and crystal arrangement

Crystal	ρ	Arrangement from ρ	observed
BeO	0.21	B-W	W
MgO	0.45	cfc	cfc
ZnO	0.48	cfc	W
CaO	0.71	cfc	cfc
SrO	0.9	bcc	cfc
PbO	0.86	bcc	layer
CaF_2	0.73	F	F
MgF_2	0.49	F	R
NiF_2	0.51	F	R

(W = wurtzite, B = zinc-blende, F = fluorite, R = rutile)

Another characteristic example is that of silicates: M_2SiO_4, where M = Be, Mg, Mn, Zn. The size of O^{2-} ions (r = 1.40 Å) is very large with respect to the radius of Si^{4+} (0.41 Å). Therefore the classical theory considers the lattice as a packing of O^{2-} ions. The Si^{4+} ions are located in the tetrahedral holes (ρ = 0.29). The difference between the coordination number of the Be atoms (4) and those of the Mg, Fe, or Mn (6) is explained by the large difference in ion size:

Be: 0.31;   Mg: 0.65;   Fe: 0.75;   Mn: 0.80 .

"The larger the ions M, the larger is the number of neighboring oxygen atoms": the beryllium atom is located in tetrahedral holes, whereas the others, which are larger, are located in octahedral holes. Nevertheless in the willemite, Zn_2SiO_4, the zinc atom, despite its size (r=0.74), possesses a coordination number of 4, and not 6.

These examples clearly show that, although the impenetrability radius of atomic inner shells plays an undeniable role in the determination of the structure, it is not the main factor.

3. Quantum Study of Relative Stabilities of Various Arrangements in Alkali Halides

The cellular model that we proposed in order to describe the structure of alkali halides allows ab initio-SCF calculations. First, one determines the molecular orbitals corresponding to various cells. Since the electron charge transfer from halogen atoms toward metal atoms is very weak, a simple perturbation calculation yields the energies of the cells in the crystal [22]. In Table 16 are listed the energies per mole MX, obtained in lithium, sodium, and potassium, fluoride and chloride. The interatomic distances used are those observed in the cfc arrangement.

From the values shown in Table 16, it can be seen that in the six halides considered here, the most stable is the cfc arrangement. It should be noted that the result was obtained assuming for the three structures (cfc, bcc, zinc blende) the same interatomic distance, the bcc experimental distance. This result clearly shows that, independently of the possible effect of the radius ratio, the crystal arrangement is also due to electronic factors.

Table 16

Energies of various arrangements in alkali halides

Crystal	c.f.c. (ν = 6)			b.c.c. (ν = 8)			blende (ν = 4)		
	E_{cell}	E_{ion}	E_{lat}	E_{cell}	E_{ion}	E_{lat}	E_{cell}	E_{ion}	E_{lat}
LiF	-36.018	-35.855	-27.054	-40.135	-40.072	-26.986	-31.929	-31.639	-26.950
NaF	-34.323	-34.205	-26.548	-37.931	-37.872	-26.497	-30.738	-30.539	-26.408
K F	-32.793	-32.710	-26.058	-35.925	-35.877	-26.046	-29.644	-29.542	-25.903
LiCℓ	-24.808	-24.645	-17.788	-28.020	-27.947	-17.722			
NaCℓ	-23.878	-23.749	-17.492	-26.807	-26.752	-17.441			
KCℓ	-22.909	-22.814	-17.186	-25.541	-25.505	-17.147			

E_{cell} = energy of the elementary cell;

E_{ion} = energy of the cell assuming that the lattice is perfectly ionic (i.e., without delocalization);

E_{lat} = energy per MX in the lattice (a.u.).

Table 17 shows the decrease in stability between the cfc and bcc arrangements in the series Li-Na-K.

Table 17

Difference in the stability between
the cfc and bcc arrangements

Crystal	$E_{cfc} - E_{bcc}$	
LiF	-0.067 a.u. =	-42.2 kcal
NaF	-0.041	-25.6
K F	-0.012	- 7.3
LiCℓ	-0.065	-40.8
NaCℓ	-0.052	-32.3
KCℓ	-0.039	-24.7

Whatever certainty that it is possible to grant to the values of the differences $E_{cfc}-E_{bcc}$, it is certain that, for a given halide in the series Li-Na-K, this difference decreases. Though we have not performed calculations for rubidium and cesium atoms, it is not unreasonable to think that this tendency would continue for these atoms and that the bcc arrangement is the most stable. That would be in agreement with experimental results.

A more detailed examination of the expression of the energy provides insight into the roles of various factors. First, the stabilization arising from the delocalization of electrons from ions X^- toward the metal atoms, decreases when the number of neighbors (ν) increases (Table 18).

Table 18

$\Delta E = E_{ion} - E_{cell}$ = energy of stabilization with respect
to the ionic model, arising from the
delocalization (in a.u.).

Crystal	$\nu = 4$	6	8
LiF	0.290	0.163	0.063
NaF	0.199	0.118	0.059
K F	0.102	0.083	0.048

This result can appear quite surprising. By analogy to the results obtained in the conjugated molecules, in which the increase of the delocalized system always corresponds to greater stabilization [7], one should expect an opposite result. In fact, the problem is complicated by the existence of an antagonist factor. The increase of ν requires the use of hybrid atomic orbitals, built up from orbitals arising from higher and higher levels: $\underline{d}$ then $\underline{f}$. The use of such hybrids corresponds to a loss of energy for the metal atoms. This loss is opposed to the gain obtained from the delocalization of electrons.

Moreover, the stability of the structure is conditioned not by the energy of the isolated cell, but by that of the cell in the crystal, i.e., exposed to the field of the other cells, which here can be replaced by point charges equal to ±1. More precisely, the energy is:

$$E = E_{cell} + \nu \frac{N}{a} - \frac{NA}{a} \tag{35}$$

where N is the number of electrons per cell (here: 8); $\underline{\nu}$, the coordination number (8, 6, or 4); $\underline{A}$, the Madelung constant of the lattice (1.7476 in cfc arrangement, 1.7627 in bcc arrangement, and 1.6381 in zinc-blende structure); $\underline{a}$, the distance between nearest neighbors.

The energy can be written as a function of the stabilization ΔE with respect to the ionic model:

$$E = \Delta E - \frac{NA}{a} \tag{36}$$

Consequently, in cfc and bcc structures, in which the Madelung constants are practically equal, the factor ΔE prevails. In contrast, in the zinc-blende structure, the Madelung constant is lower and the stability arising from the delocalization is destroyed by the electrostatic term, which is too weak in absolute value.

In a purely electrostatic model, where only the Madelung energy (-NA/a) would be introduced, one could predict that the bcc arrangement would be the most stable.

This short discussion clearly shows the complexity of the problem. The choice of the crystal arrangement cannot be reduced to only one factor. The structures arise from a balance of many factors.

4. Cohesive Energy and Crystal Arrangement in Metals

Besides perfectly ionic lattices, it is particularly interesting to study crystal arrangements in metals. Nevertheless, we have seen the difficulties involved in the calculation of the cohesive energy. In the case of metals with one valence electron (alkaline in particular), the simplified formula (22) gives a good value for the cohesive energy. When the crystal arrangement depends on this energy, we can assume that formula (22) is sufficient for a qualitative discussion.

Metals with many electrons will be discussed further.

We shall neglect linear metal chains, which have been the subject of many publications, but which in fact do not constitute a very realistic case.

5. Why is Lithium a Metal?

This question is the first which must be asked. The answer will be a test for the approximate method, based on relation (22).

Lithium crystallizes in the bcc system. The distance between nearest neighbors is 3.04 Å [16]. The cohesive energy per atom is -1.65 eV [33].

After (22):

$$\varepsilon_{bcc} = 8\ell\, \beta_{bcc} + 2\ell^2\, J_{bcc} \tag{37}$$

($\nu = 8$). According to reference 50, the bond order ℓ is equal to 0.26. To calculate the coulombian integral J, we shall use an Ohno-type [67] formula, which very quickly leads to sufficiently correct values [68]:

$$J(r) = \frac{27.20}{\sqrt{A + r^2}}$$

(J in eV, r in atomic unity).

For $r = 0$, $J = \mathcal{J} - \mathcal{A} + \varepsilon(\mathcal{J} + \mathcal{A})$ (ref.69). Now $\mathcal{J} = 5.39$ eV and $\mathcal{A} = 0.54$ eV [32]. Consequently:

$$J\ (r=0) = 3.3 \text{ eV} \quad \text{and} \quad A = 68.$$

Formula (22) gives:

$$\beta_{bcc} = -0.98 \text{ eV}$$

If the crystal were a molecular crystal, i.e., composed of Li_2 molecules, the bond order between two neighboring atoms would be: 1, if the atoms belong to the same molecular Li_2; or null, in the other case. The interatomic distance $r(Li_2)$ would be 2.67 Å as in the isolated molecules. Consequently, the cohesive energy per atom ($\nu = 2$) would be:

$$\varepsilon_{mol} = \beta_{mol} + 0.25 \ J_{mol} \qquad (39)$$

In our simplified method, ε_{mol} is equal to half of the dissociation energy of the Li_2 molecule (-0.56 eV). Since after Eq. (38): $J(2.67) = 2.81$ eV, we obtain:

$$\beta_{mol} = -1.26 \ eV$$

This value is an absolute value, higher than that corresponding to the bcc crystal (-0.98). The ratio β_{bcc}/β_{cfc} is equal to 0.78.

We indicated the usual approximation, which assumes that the β are proportional to the corresponding overlap integrals S [31]. In this circumstance, the ratio β_{bcc}/β_{cfc} would be 0.504/0.585 = 0.86. With this last value, we obtain:

$$\beta_{mol} = -1.14 \ eV \quad and \quad \varepsilon_{mol} = -1.26 \ eV$$

The molecular crystal is always less stable than the metallic lattice, but the value obtained for ε_{mol} would be almost twice the experimental value. We conclude that the decrease of the β integrals with respect to the distance is in fact faster than is that of S integrals.

However, this discussion clearly shows that the approximate formula (22) corrected by hypothesis (24), leads to very satisfactory results.

6. Why Does Lithium Adopt the bcc Arrangement?

The previous calculation shows that lithium adopts a metal structure. The cohesive energy of the bcc lattice is higher in absolute value than that of a molecular lattice. However, other arrangements are possible.

First, let us examine the cfc arrangement.

Experiments show that the ratio of distances between nearest neighboring atoms increases by about 3% from the bcc system ($\nu = 8$) to the cfc system ($\nu = 12$). Consequently, in a hypothetical cfc lattice of lithium, the distance would be 3.16 Å.

We would have for such a lattice:

$$\varepsilon_{cfc} = 12\ell\ \beta_{cfc} + 3\ell^2\ J_{cfc} \qquad (40)$$

Assuming the proportionality law between β and S:

$$\beta_{cfc} = 0.948\ \beta_{bcc} = -0.93\ eV.$$

After [70], in the cfc lattice, ℓ can be estimated as being close to 0.15. Assuming the value to be 0.16, which favors stability and the cfc lattice, we obtain:

$$\varepsilon_{cfc} = -1.58\ eV$$

After the remark related to the inadequacy of law (24) ($\beta = k\alpha S$), the value (-1.58) obtained for the cfc lattice is certainly too high in absolute value. Nevertheless, this value is higher than that corresponding to the bcc lattice (-1.65). Consequently the bcc arrangement is more stable than the cfc arrangement.

7. Why does Cubium not exist?

It is surprising that the simple cubic lattice (cubium) does not exist . This fact can be explained without difficulties in the case of lithium.

In such a lattice, $\ell = 0.33$ [50]. The coordination number is equal to six. Consequently:

$$\varepsilon_{sc} = 2.0\ \beta_{sc} + 0.17\ J_{sc} \qquad (41)$$

In order to estimate ε_{sc}, it is necessary to know the interatomic distance in the lattice. In the absence of experimental values, we shall adopt the following empirical law, which cannot satisfactorily relate the distance to the bond order:

$$r(\overset{\circ}{A}) = \frac{3.51 + 6.05\ell}{1 + 2.58\ell} \qquad (42)$$

For $\ell = 1$, one obtains 2.67 Å (molecule Li_2); for $\ell = 0.26$, 3.04 (bcc lattice); for $\ell = 0.16$, 3.16 (cfc lattice).

The interatomic distance corresponding to $\ell = 0$, is simply the distance between second-neighbors in the bcc lattice.

From relation (42) one deduces that in a simple cubic lattice, where $\ell = 0.33$, the distance between nearest neighbors is about 2.97. The corresponding overlap integral S is equal to 0.518. Consequently, $J_{sc} = 2.73$ eV and $\beta = -1.01$ eV. Whence we conclude:

$$\varepsilon_{sc} = -1.56 \text{ eV}$$

This value shows that the simple cubic lattice would be less stable than the cfc lattice, but more stable than the molecular crystal.

8. Hydrogen

Under normal conditions, solid hydrogen is a molecular crystal. A priori, one could assume that, owing to its electronic structure, similar to that of alkaline metals, the solid hydrogen is a metal. In fact, we saw that low-molecular-weight compounds are inclined to give molecular lattices. Nevertheless, it is likely that, at sufficiently high pressure ($\sim 3 \cdot 10^6$ bar), hydrogen would behave as a metal. Many sophisticated theoretical studies have been performed on the metallic hydrogen problem [71]. Here we are content with the simple analysis that we used with success in the lithium case.

In the molecular crystal, we have:

$$\varepsilon_{mol} = \beta_{mol} + 0.25 \ J_{mol} = -2.34 \text{ eV} \tag{43}$$

The H_2 distance is equal to 0.74 Å. In the hydrogen atom $\mathfrak{J} = 13.60$ and $\mathcal{A} = 0.65$ eV.
Consequently [68]: $J(r=0) = 8.4$ eV.
Whence:

$$J(r) = \frac{27.20}{\sqrt{10.5+r^2}} \tag{44}$$

which leads to $J(0.74) = 7.7$ eV. Consequently:

$$\beta_{mol} = -4.3 \text{ eV}$$

In the bcc lattice, we should have:

$$\varepsilon_{bcc} = 2.06\ \beta_{bcc} + 0.133\ J_{bcc} \tag{45}$$

The problem is to determine the distance between nearest neighbors. The value in general adopted for the Van der Waals radius of hydrogen [7] is 1.1 ± 0.1 Å. In the linear system H_4, one can consider that we have two H_2 molecules, practically independent [72]. The distance between the inner hydrogens is ~ 2.2 Å. Consequently, one can approximate that a bond order ℓ equal to zero, corresponds to a distance at least equal to 2.2 ± 0.2 Å. Assuming a linear law of variation between the bond order and the interatomic distance in the range $r = 0.74$ Å (H_2 molecule) to 2.2 Å, we obtain:

$$r\ (\text{Å}) \sim 2.20 - 1.56\ \ell \tag{46}$$

For $\ell = 0.26$, in the hypothetical bcc lattice, $r \sim 1.82$ Å.

Consequently, $J = 5.8$ eV. Assuming that β is proportional to S, $\beta(1.82) \sim -1.5$ eV and $\varepsilon_{bcc} = 2.3$ eV.

In fact, the decrease of β is faster. Certainly $|\beta_{bcc}| < 2.3$ eV. We conclude that the solid hydrogen is a molecular crystal.

Relation (46) shows that ε_{bcc} is lower than ε_{mol} for a distance of about 1.6 Å. That would explain the metallic properties of hydrogen at high pressure, independently of any introduction of p orbitals.

9. Copper, Silver, and Gold

The electron structures of copper, silver, and gold are similar to those of the alkaline metals. Practically, one can consider that each atom brings only one s electron to the lattice.

In copper [32], $\mathcal{J} = 7.7$ eV and $\mathcal{H} = 1.2$ eV. Consequently $J_o = 5.67$ eV and

$$J(r) = \frac{27.2}{\sqrt{23+r^2}} \tag{47}$$

In the cfc lattice, the interatomic distance is equal to 2.56 Å. In the bcc lattice, this distance would be 2.48 Å. Relations (37) and (40) give:

$$\varepsilon_{cfc} = 1.9\ \beta_{cfc} + 0.3$$

$$\varepsilon_{bcc} = 2.1\ \beta_{bcc} + 1.4$$

The experimental cohesive energy, corresponding to the cfc lattice, is equal to -3.5 eV [33]. Consequently, β_{bcc} = -2.0 eV. In lithium the ratio β_{cfc}/β_{bcc} is about 1.05. Here, the value of this ratio is lower because the interatomic distances are greater. Consequently $|\beta_{cfc}| < 2.1$ eV and $|\varepsilon_{bcc}| < 3$ eV. This value $(<|\varepsilon_{cfc}|)$ explains why the copper adopts the cfc arrangement.

10. Alkaline-Earth Metals and Semi-Metals

In the alkaline-earth metals, each atom brings two electrons. A calculation that introduces only the s orbitals gives doubly used molecular orbitals. The cohesive energy would be equal to zero and the crystal would be an insulator. In reality, the p levels, which are not used in the ground atomic state, give a p band which overlaps the top of the s band.

Consequently, the molecular orbitals coming from the s atomic orbitals (s band) are not all doubly used, and the lowest molecular orbitals coming from p atomic orbitals are doubly used.

To describe the structure of these metals, it is necessary to introduce simultaneously the s and p atomic orbitals. The structure differs from that in alkaline metals or in the copper series: the arrangement is hexagonal close-packed.

When the metal brings a higher number of electrons to the delocalized system, as in bismuth, the molecular orbitals coming from the s atomic orbitals are doubly used. The structure is imposed by the p orbital system. It explains why the bismuth lattice is very close to a simple cubic lattice [73]. The crystal is composed of infinite p orbital chains, in three orthogonal directions. Each bismuth atom brings three electrons to this system; the treatment of each chain is formally identical to that in a polyen:

It is known that in such molecules the bonds are alternatively short and long [7]. The situation is the same in bismuth. There are layers composed of hexagonal cycles (form <u>chair</u>, with angles of 90°). These layers are joined by longer and consequently weaker bonds.

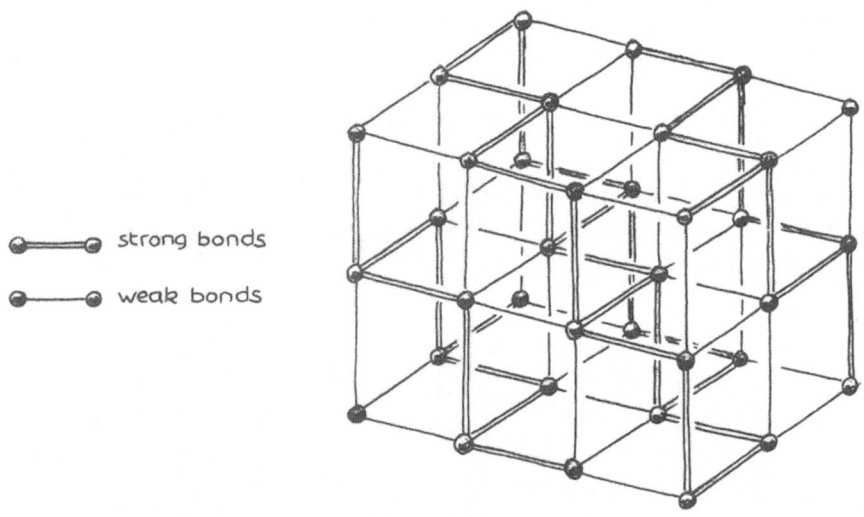

11. <u>Transition Metals</u>

We have seen that in the transition metals, the <u>d</u> orbitals play the main role. The <u>a priori</u> determination of the structure is more complex. We shall cast this problem aside (see, e.g., [74]).

12. <u>Crystal Arrangements Allowing a Description in Terms of Localized</u>
    <u>Orbitals</u>

In such crystals, the structure follows from the possibility of building hybrid atomic orbitals pointing rigorously at one another or with a small angular deviation.

The case of elements belonging to the carbon column (Si, Ge, ...) or the $A^{III}B^{V}$ compounds (BN, BeO, GaAs, ...) is particularly simple.

The symmetric sp_3 hybridization allows a regular occupation of the space without distortion. The atoms arrange themselves in cycles of six atoms. The two sole possibilities for these cycles are the form <u>chair</u> and the form <u>boat</u>. With only <u>chair</u> cycles, we have the zinc-blende structure (cubic system), with <u>chair</u> and <u>boat</u> cycles we have the wurtzite structure (hexagonal system). Likewise, in compounds in which a $sp_3 \times sp_3 d_3 f$ hybridization is possible, the fluorite arrangement is the sole possible arrangement (cubic system).

The problem lies in the prediction of the hybridization. In general, the hybridization can be derived from the chemical formula. To illustrate this research, we shall examine in detail the case of oxides.

13. Oxides Arrangements

The general formula is $M_x O_y$. First, we shall consider the case where all the atoms M and all the atoms O are equivalent. Let ν_O and ν_M be the number of localized orbitals, respectively, carried by the atoms O and M. We have:

$$\nu_M = \frac{x}{y} \nu_O \qquad (48)$$

In general, the coordination number of the oxygen atom does not exceed 4. A higher coordination number would involve the use of orbitals $\underline{d}$ or $\underline{f}$, which would destabilize the system.

Examine the various cases that can occur:

1. MO   (x=y=1):

    (a) $\nu = 1$: the MO groups are independent. We have a molecular crystal (e.g.: CO)

    (b) $\nu = 2$: It would have a zigzag chain structure if the atom M carries two nonbonding used orbitals (e.g., sulfur atom):

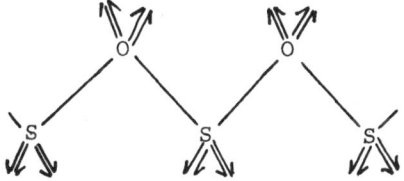

Such a system, which would possess only simple S-O bonds, would be less stable than the conjugated molecular system:

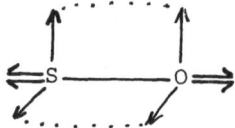

That explains why it does not exist.

An other possibility would be, with alkaline-earth metals, to build linear chains:

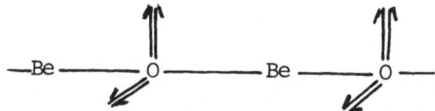

The delocalization of the lone pairs of the oxygen atoms is always very weak [7]. The stabilization that would result from the formation of two conjugated systems ($\pi + \pi'$), as in acetylenic compounds [7], will be very weak. The system prefers to adopt a structure that allows the formation of simple bonds in higher numbers, which will increase the stability.

(c) $\nu = 3$: We would have layers as in BN (graphite structure). However the oxygen atom would necessarily be positive and the M, negative. That would be very unfavorable. This structure does not exist.

(d) $\nu = 4$: The oxygen atom is sp_3-hybridized. Various possibilities exist for the metal hybridization.

(α) M is sp_3-hybridized: In this case, the crystal adopts the zinc-blende or the wurtzite structure. In fact, it seems that the wurtzite structure is more frequent: BeO, ZnO.

(β) M is sp_2d-hybridized: The four hybrids carried by the metal point into four directions, in the same plane. It is impossible to obtain an alignment of the sp_3-hybrids of oxygens that point in directions separated by angles of 109.5°. Consequently, the packing will be obtained from bent orbitals. The lattice will be compressed (quadratic system).

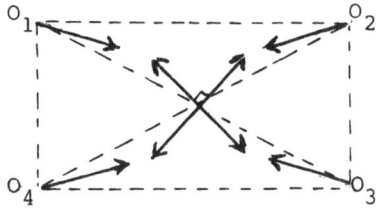

E.g., PdO, $\alpha = \overparen{O_1 Pd\ O_2} = 98°$. In PdS, the distance PdS is
larger than PdO, the default of alignment is less weak:
$\alpha \sim 90°$.

    In tenorite CuO, the situation is similar. But the
length of the CuO bonds is less than with the platinum or
the palladium. The distortion is greater. The oxygen atoms
are located in the corners of a parallelogram. The system
is triclinic.

(γ) The hybridization of M is greater, but the metal does
not use all the hybrids in order to build the bond orbitals.
E.g., PbO. The lead atom is sp_3d-hybridized ($d = d_{xy}$) but
it carries nonbonding doubly used orbitals. We have layers.

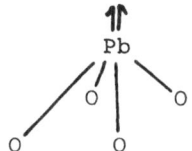

2. MO_2 ($\nu_M = 2\ \nu_O$)

(α) $\nu_O = 1$, $\nu_M = 2$. We have isolated entities, MO_2. The
crystal is a molecular crystal. M can be hybridized in sp
(CO_2) or in sp_2 (SO_2).

(β) $\nu_O = 2$, $\nu_M = 4$ (sp_3 hybridization). We have a
tridimensional lattice. Various possibilities occur
depending on the hybridization of the oxygen atom.
→ sp: cristobalite, tridymite. The lattice possesses the
symmetry of the silicon lattice from which it is derived
(zinc-blende or wurtzite).
→ sp_3: quartz, GeO_2. We have helices; right-handed, or
left-handed, which explaines the rotatory power of these
crystals.

(γ) $\nu_O = 3$, $\nu_M = 6$ (sp$_3$d$_2$ hybridization)

E.g., TiO$_2$ (rutile), GeO$_2$, SnO$_2$, PbO$_2$.

The alignment of the axes of metal orbitals with those of oxygen atoms that point in directions separated by angles of 120°, causes compression of the lattice (quadratic system).

(δ) $\nu_O = 4$, $\nu_M = 8$ (sp$_3$d$_3$f hybridization). The alignment of metal orbitals with those of oxygen is possible. We have a cubic system (fluorite structure). E.g.: ThO$_2$, HfO$_2$. The coordination number of the metal increases when the atom becomes larger. That is logical because the excited orbitals $\underline{d}$ or $\underline{f}$ are more and more easy to reach. That explains the increasing values of ν_M (and consequently of $\nu_O = \nu_M/2$).

3. MO$_3$ ($\nu_M = 3 \nu_O$)

(α) $\nu_O = 1$, $\nu_M = 3$. The crystal is a molecular crystal: SO$_3$.

(β) $\nu_O = 2$, $\nu_M = 6$ (sp$_3$d$_2$ hybridization)

→ O is sp-hybridized: ReO$_3$ (cubic lattice)
→ O is sp$_3$-hybridized: MoO$_3$ (double layers, the Mo atoms are located in two parallel planes).

(γ) $\nu_O = 3$, $\nu_M = 9$. The equivalence between atoms of the same kind is not respected. Regular polyhedra with 9 faces or 9 apexes do not exist. We shall eliminate this case.

4. MO$_4$ ($\nu_M = 4 \nu_O$)

(α) $\nu_O = 1$, $\nu_M = 4$. We have a molecular crystal. E.g.,: OsO$_4$.

(β) $\nu_O = 1$, $\nu_M = 8, 12, \ldots$ The cases are imaginable. In fact, they do not exist. Such structures would be obtained, e.g., from metallic lattices bcc ($\nu = 8$) or cfc ($\nu = 12$), in which an oxygen atom would be located between two M atoms (cf. cristobalite). However, the size of the oxygen atoms with respect to that of metal atoms forbids such a packing.

5. M$_2$O ($\nu_M = \nu_O/2$)

(α) $\nu_O = 2$, $\nu_M = 1$: the crystal is a molecular crystal.

E.g., H_2O.

(β) $\nu_O = 4$, $\nu_M = 2$: the sp_3-hybridized oxygen atoms give a diamond-like lattice. Between each pair of oxygen atoms, there is a sp-hybridized metal atom. The structure is that of a cristobalite in which the roles of oxygen and silicon atoms are inverted. E.g.: Cu_2O, Ag_2O. The system is cubic.

(γ) $\nu_O = 8$, $\nu_M = 4$: antifluorite structure (cubic system). E.g., Li_2O, Na_2O, Rb_2O.
In this structure the hybridization of the oxygen atom is unusually high. Such a hybridization is possible owing to the large gain in energy that arises from the high electronegativity difference between oxygen and an alkaline metal. These crystals are very brittle due to the high hybridization. Hydrolysis will be very easy, in contrast to Cu_2O or Ag_2O, which are insoluble in water.

6. M_2O_3 ($\nu_M = 3\,\nu_O/2$)

(α) $\nu_O = 2$, $\nu_M = 3$. We have chains. E.g.: Sb_2O_3.

(β) $\nu_O = 4$, $\nu_M = 6$.

The coordination number 6 of the metal corresponds to an sp_3d_2 hybridization ($A\ell_2O_3$, Fe_2O_3, ...) or to a higher hybridization (sp_3d_3f in Bi_2O_3 or in rare earth oxides with nonbonding orbitals). The structure involving nonbonding orbitals appears when the number of electrons is too large to build eight binding orbitals, as well as in heavy elements in which the orbitals are easy to reach. The structure will be hexagonal if the hybridization is sp_3d_2 (corundum). If the hybridization is sp_3d_3f, the structure will be that of fluorite, but with certain atoms absent (Mn_2O_3). The lattice will be practically cubic.

7. M_xO_y (x > 2): We have molecular crystals.

E.g., P_4O_6 ...

If all oxygen atoms are not equivalent, the situation is more complicated. Nevertheless, enumeration of the cases is possible. Assume that all the M atoms are equivalent but that there are two kinds of oxygen atoms. In MO_2, e.g., the structures are those obtained for MO with an additional oxygen atom, linked to each M atom. E.g., SeO_2.

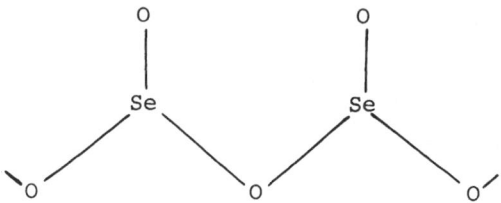

14. Role of Hybridization

The example of oxides has shown the large role played by the hybridization in the choice of structure. This phenomenon is a general one and must be studied with great care because it conditions at once the choice of the lattice (as in MO_2) and sometimes its stability (as in MO_2).

A striking example is that of the silicates M_2SiO_4, built from SiO_4 groups joined by metal atoms. $\underline{M}$ can be beryllium, magnesium, iron, or manganese, and the use of $\underline{d}$ orbitals allows sp_3d_2 hybridization ($\nu = 6$). For zinc, the $\underline{d}$ orbitals are all used and do not intervene in order to build the molecular bonding orbitals. Consequently the zinc atom will be sp_3-hybridized and the coordination number will be 4. The zinc atom will be more analogous with the beryllium atom than with the magnesium. Despite its large ionic radius (0.74 Å), the zinc atom adopts a coordination number of 4 in willemite, whereas the magnesium atom, whose radius is smaller (0.65 Å), possesses in phenacite a coordination number of 6. For zirconium, a coordination number of 8 is possible because the sp_3d_4 hybridization is easily accessible. This hybridization will be more easily reached by heavier atoms: thorium and uranium, which will replace zirconium atoms in the altered variety of zircon-the so-called malacon. The sp_3d_4 hybridization imposes a quadratic symmetry.

We have seen that in most ionic crystals, the coordination number is, in general, higher than that in other crystals. This would explain that in $ZnCO_3$, the hybridization of the Zn atom is sp_3d_2. The carbonates are more ionic than the silicates (e.g., $Mg_2SiO_4/MgCO_3$). The same is true for ZnF_2.

We see clearly that, though the increase in atomic size corresponds directly to an increase in the coordination number, the origin of the phenomenon does not lie in the geometric conditions but rather in the possibility for the atom to have access to excited states. In the classical interpretation, one says that the atoms try to be surrounded by the largest number of atoms, and that the geometric conditions limit this number. The larger the atom, the more neighbors there will be of a given kind. In fact, we see that the larger the atom, the more possibilities it has of possessing an electron structure that permits the building up of a larger number of molecular bonding orbitals, owing to the larger number of hybrids which the atom can carry. This does not mean that the size of the atom does not play a role. The size can intervene to impose geometric constraints, as we shall see in further examples. However, these constraints will never be the main factor.

An interesting situation, which will illustrate these considerations, arises in carbides, MC_2. We have seen the structure of calcium carbide, CaC_2. The calcium atoms are sp_3d_2-hybridized. Each calcium is linked to two C_2 groups by the terminal carbon atom of each group and to four C_2 groups by their π-orbitals. For thorium, the highest hybridization is possible, though the size of this atom is practically the same as that of calcium. The number of orbitals linked to carbon atoms of neighboring C_2 groups can be higher.

Consequently, we have a rotation of C_2 groups.

A sp_3d_2 hybridization is not possible for the beryllium atom. A CaC_2-like structure does not exist. The formula of carbide is Be_2C. The structure is the antifluorite structure. The very unusual sp_3d_2f hybridization of carbon explains the very easy hydrolysis, which gives CH_4. The other MC_2 carbides, which possess C_2 groups, give acetylen, C_2H_2.

15. Importance of the Alignment of the Orbitals and Effects of Distortions

With regard to oxides, we have seen that the impossibility of aligning the axes of the atomic orbitals by twos, causes distortions (PtO, TiO_2), which explains the choice of the lattice by the crystal. This phenomenon is general. Whenever the alignment is possible, it is realized. If this condition is not possible, the packing will nevertheless be obtained, insofar as the default of alignment remains weak.

The alignment of the orbitals is obtained without difficulty for rigid orbital systems: e.g., sp_3 hybrids as in zinc-blende or wurtzite. In phenacite-like crystals, M_2AO_4, the alignment occurs between the sp_3-hybrids of M and the sp_3-hybrids of one oxygen of the

AO_4 neighboring groups. The structure is rhombohedric ($\alpha \sim 108°$).

If alignment of the orbital axes is impossible, various cases can occur.

For example, in zircon, owing to the strong repulsion between the electron charges carried by the nonbonding hybrid orbitals of the oxygen atoms in SiO_4 groups, the sp_3-hybridization of the oxygen atoms is unsymmetrical. The angle between any two hybrid directions (Δ_i, Δ_j) is greater than 109.5°, but lower than 120° (symmetrical sp_2--hybridization). A value about 116° seems very plausible. The $\widehat{SiO\Delta_i}$ angles are equal to about 102°. The center of gravity of the oxygen hybrid orbital is located at about 0.3 Å from the oxygen nucleus [9]. The oxygen hybrid orbitals form a rigid system. The zirconium atom is sp_3d_4-hybridized. The atomic orbital directions pointing toward the oxygen atoms of SiO_4 groups located whether above or under the zirconium atom, form angles of about 65° (see Appendix B). The maximum of stability is obtained when the zirconium hybrids point toward the centers of gravity of the oxygen hybrids. An elementary calculation gives for the Zr...Si distance the value 3.1 Å. The experimental value is c/2 = 3.0 Å.

As another example of weak distortion, we shall quote the case of cryolithe, Na_3AlF_6.

The simplest structure that can be imagined is a network of octahedra, AlF_6, located in nodes of a bcc lattice and joined by sodium atoms. In fact, at high temperature, cryolithe possesses a cubic structure. The sp_3d_2-hybrid orbitals of sodium point toward orbitals carried by fluorine atoms and pointing from F to Al.

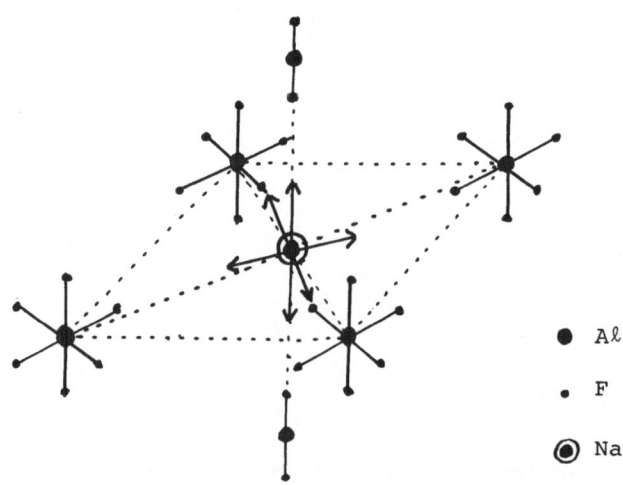

● Al

. F

◉ Na

At room temperature, the alignment defect is about 15°. Nevertheless the distortion remains sufficiently weak so as not too much have modifying effect on the interatomic distances. Although in theory the lattice is monoclinic, practically the arrangement is cubic.

In contrast, the distortion effect is greater in potassium chloroplatinite (K_2PtCl_4).

The potassium atom is surrounded by eight chlorine atoms, located at equal distances. A sp_3d_3f hybridization is not possible. The chlorine atoms carry sp-hybrid orbitals whose axes are the Pt-Cl lines. Alignment with the potassium orbital is not possible. The centers of charges of the chlorine hybrids are located at the corners of a square. A sp_3d_3f hybridization will require a too weak interplanar distance c, which is incompatible with the presence of doubly used nonbonding orbitals carried by platinum atoms (high electron density corresponding to the classical lone pairs). Consequently the distance a between platinum atoms will be larger than c. The system is quadratic.

A p_3d_4f hybridization with orbitals pointing toward the centers of charge of the chlorine sp-orbitals corresponds to a vertical drawing. Such a hybridization is possible only for heavy atoms (e.g., K), because it requires high energies.

Let us M the intersection points of the PtCl directions and the axis of the hybrid orbitals carried by the potassium atom. The points M form a square, whose the edge (a/2) is equal to 3.50 Å [65], i.e., Pt - M = 2.48 Å. The interatomic distances Pt - Cl are equal to 2.32 Å [65]. Consequently, the Cl - K bonds are weakly bend. The angle between the potassium hybrid and the vertical line is about 50° (cf. Appendix B).

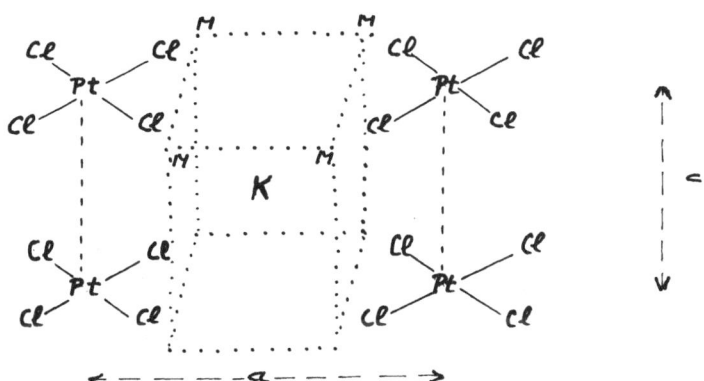

In K_2PdCl_4, the situation is identical:

$$d \ (PdCl) = d \ (PtCl) \quad \text{and} \quad c = 4.10 \ \overset{\circ}{A} \ ^{65}.$$

The structure of calcite-like MCO_3 crystals (where M = Ca, Mg, Fe, Zn, ...) can be explained in a similar manner. An elementary calculation shows that the axes of the six hybrid orbitals sp_3d_2 of the metal atom, intersect the corresponding lines CO in points P located at about the same distance (0.2 Å) from the oxygen nucleus. The center of gravity G of the sp-hybrid carried by the oxygen atom is located approximately at OG = 0.4 Å. This shows that the hybrid orbitals carried by M and the sp-orbitals of the corresponding oxygens overlap each other, so as to allow the formation of bent localized molecular orbitals.

In part II, we have given the corundum structure. Often one considers that the oxygen atoms form a slightly distorted hexagonal close-packing lattice with metal atoms in some of octahedral holes. In fact, in a perfect hexagonal close-packing oxygen lattice, the four aluminum atoms linked to an oxygen atom would form with this atom a square pyramid having the oxygen atom as the apex. This disposition would correspond to a sp_3d hybridization (see Appendix B) which requires a too high energy. For the oxygen atom, a sp_3 hybridization is more favourable. Nevertheless, the four sp_3 hybridized orbitals cannot point at four directions located in a same half of the space.

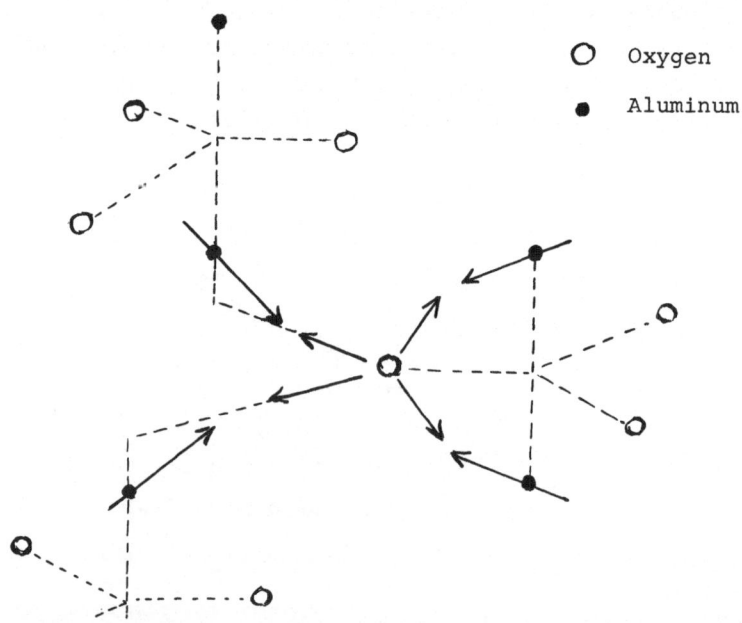

○ Oxygen

● Aluminum

Another description considers corundum crystal as built up from Al_2O_3 groups (Fig.4, p.33). In fact, the interatomic distances Al-O in Al_2O_3 groups are larger (1.97 Å) than those between one aluminum atom and one oxygen atom, belonging to two neighboring Al_2O_3 groups (1.86 Å). The actual structure is detailed in the above scheme. The sp_3-hybrid orbital of the oxygen atoms do not point toward the aluminum atoms. We have a tridimensional macromolecular lattice with bent bonds.

16. Conjugation Effect in Many Crystals

In molecules, the conjugation of unsaturated systems, i.e., possessing molecular orbitals built up from π-atomic orbitals (2p-orbitals in hydrocarbons: benzene), is, in general, a planarity factor [7] (e.g., butadiene). Nevertheless, steric hindering can oppose this tendency to make a planar system (e.g., biphenyl, cyclooctatetraen [7]). The same phenomena are encountered in crystals.

Consider sodium nitrate. By analogy with sodium chloride, one would expect a cfc lattice, the NO_3^- ions replacing the Cl^- ions, and possessing arbitrary orientations in space. The nitrate ion possesses a π-conjugated system which imposes the planarity [7]:

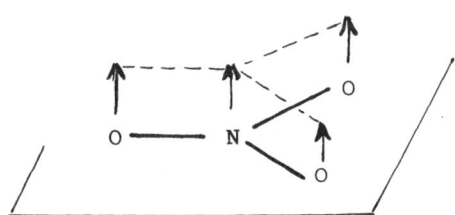

Rather than maintaining arbitrary orientations in space, the NO_3^- groups arrange themselves parallel to one another so as to make conjugated planes. The energy of the π-system is minimum. The repulsion between the oxygen atoms of nitrate groups, caused by the sizes of the atoms and their negative net charge, provokes a weak removal of NO_3 groups, which compresses the cubic lattice into a rhombohedric lattice.

The same explanation is valid for calcite where the carbonate ion has a structure similar to that of the nitrate ion.

In N_3M azides, the N_3^- groups are linear. The structure of the N_3^- ion is:

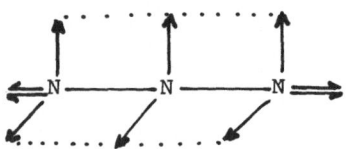

We have seen that N_3K possesses a quadratic structure derived
from that of NaCℓ. The N_3 groups are independent of one another. In
N_3Na, the distance between the central atoms of the neighboring N_3
groups is less than in N_3K. A general rotation of the N_3 groups,
whereby these groups are arranged parallel to one another, allows
the formation of an infinite conjugated system, built up from linear
small stick N_3. The lattice is deformed into a vertically strained
rhombohedron. In $NaNO_3$, the rhombohedron is compressed. Such an
arrangement is not possible with potassium atoms because the distance
between the N_3s would be too large to allow the conjugation to effect
sufficient stabilization.

Finally, let us recall the role of conjugation in molecular
crystals. Conjugation imposes the respective disposition of conjugated
molecules (benzene, naphtalene, ...).

In the benzene crystal (p.20), the rings are not packed one upon
the other, as one would expect. The rings are located obliquely in
alternate rows. Such an arrangement is due to the conjugation. In
order to show that, let us consider the simple case of two ethylene
molecules, located in parallel planes, in the two following positions:

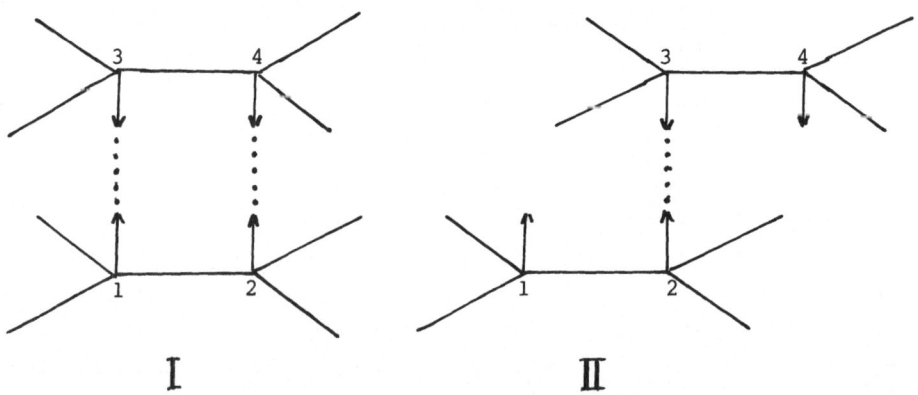

In the first position (I), the axes of 2p-orbitals coincide with
one another. In the second scheme, the axes of orbitals carried by
atoms 2 and 3 are practically aligned, but the other two do not

coincide. Let us calculate, using the Hückel method, the π-energy of the whole system constituted by the two molecules.

In I, the π-energy is $2 (2\alpha + 2\beta_{12})$. It is equal to one of the two isolated molecules. $\beta_{13} = \beta_{14}$ does not intervene. In II, the π-energy is:

$$2 (2\alpha + \beta_{12} \sqrt{4 + \lambda^2}) \sim 2 (2\alpha + 2\beta_{12}) + \lambda^2 \beta_{12}/2$$

where $\beta_{23} = \lambda \beta_{12}$ (λ is assumed to be small).

Consequently, the staggered arrangement in II is more stable.

An equivalent result can be found immediately for benzene. The stabilization energy between two molecules arranged as in scheme I, is equal to zero, whereas it has a value different from zero when the systems are staggered.

V. Distortions Caused by the Existence of Faces
and the Finite Sizes of Crystals

The existence of faces terminating real crystals breaks the
lattice periodicity. It brings about relatively important perturbations
in the neighborhood of the faces. In the present text, we shall leave
aside the surface-state problem and consider only geometric
modifications.

1. Classic Results and a Few Recent Experimental Facts

The relaxation effects near the faces of a sodium chloride
crystal were predicted long ago. Classical calculations of ion-ion
interactions, taking into account the Born repulsion term, suggest a
0.3% contraction between the first and the second layer [75]. Inclusion
of ion polarization [76] suggests 5%. In 1923, Zwicky [77] suggested that
the distances between nearest neighbors on the surface must be about
5% shorter than the corresponding distance in the bulk, so that the
sodium chloride crystal should have cracks on its surface. Recent
experimental studies show that in fact the situation is more
complicated. In LiF, not only is the distance between the planes of
a pair lower by 7% on the surface than in the bulk [78], but the
lithium atoms shift to the inside by 0.14 Å.

The development of LEED techniques has made it possible to obtain
interesting results in various metals (Aℓ, Ni, Cu, Zn, Mo, Ag, W).
In general, the atomic geometry of their surfaces is close (within
5%) to that of the bulk solid, except that contractions of the upper,
larger spacing appear possible for the non-close-packed faces [79].

In diamond [80], surface relaxation is less important than in
silicon [81]. This is probably related to the greater rigidity of the
diamond lattice.

In gallium arsenide, relaxation affects mainly the surface layer
[82]. The surface arsenic atoms of the (110) face are raised slightly
with respect to the surface plane, whereas the gallium atoms are
shifted to the inside. The same conclusions have been drawn for ZnO
and ZnS [83].

Finally, in molecular crystals, and in particular in rare-gas
crystals, classical calculations from the intermolecular (or, for
rare gases, interatomic) potentials predict an increase in the
distance between the surface elements and those lying immediately

below [20-84].

2. Examples of Quantum Treatments

The methods for a theoretical study of surface relaxation phenomena depend on the nature of the crystals under consideration.

a) localizable-orbital crystals (insulators and semi-conductors).

The simplest case is that of diamond and silicon. In the absence of distortions, the atoms situated on a (111) face would each carry a sp_3 hybrid orbital, with its axis perpendicular to the face and not used in localized orbitals associated with bonds between the atoms: the free hybrid orbitals are dangling orbitals. These orbitals are parallel to each other so that they form a delocalized system extending over the whole surface, much like that of graphite. However, the overlap integral associated with two neighboring dangling orbitals is negligible (0.04) compared to the integral over orbitals localized between the two atoms (0.65). A surface atom is thus subjected to a total force that tends to move it to the interior. The new geometric configuration may be determined by the maximum-overlap principle, which, although approximate, produces fairly good results for molecules [85].

In the case of localized orbitals, if the crystal is not polar, the total energy may be written as the sum of contributions from the individual molecular orbitals $\lbrack$Eq. (14)$\rbrack$. In an elementary theory of the Hückel type, these energies have the form

$$\varepsilon_{tt'} = \alpha_t + \alpha_{t'} + 2\beta_{tt'} \qquad (49)$$

t and t' being the hybrid orbitals that point toward one another. We have seen that $\beta_{tt'}$ was quite well proportional to the overlap integral $\langle t \mid t' \rangle$. In a given atom, the sum of the α_t terms is independent of the precise structure of the hybrids. In fact, for an sp_3 hybrid, one can write [7,9]:

$$t = as + b\,p_x + c\,p_y + d\,p_z = as + \sqrt{1-a^2}\,p \qquad (50)$$

p being an orbital of the same nature as p_x, p_y, and p_z, but pointing in the direction (b, c, d). Hence:

$$\alpha_t = a^2\,\alpha_s + (1-a^2)\,\alpha_p \qquad (51)$$

$$\sum_t \alpha_t = \alpha_s \sum a^2 + \alpha_p \sum (1-a^2) = \alpha_s + 3\alpha_p = Cte \qquad (52)$$

$(\sum a^2 = 1)$.

The total lattice energy is thus approximately proportional to the sum of the overlap integrals over the different hybrid orbital pairs forming the localized molecular orbitals. The most stable system will be obtained when that sum is highest.

The change in position of the nuclei involves a change in the interatomic distances, which in turn enter into the computation of the overlap integrals. A simple procedure for estimating those changes in length consists of connecting bond lengths with hybrid characters. The form of a hybrid depends on the hybridization coefficients; in the case of an $s^x p$ hybrid it will depend only on one coefficient [7]. In general, one can write

$$t = \frac{s + \sqrt{n}\, p}{\sqrt{n+1}} \qquad (53)$$

n being the hybridization degree.

Experimental data show that interatomic distances depend on these hybridization degrees. Some authors even postulate [86] a law of the type

$$r_{ij} = a - b \, \exp\left[-(n_i + n_j)\right] \qquad (54)$$

where i and j are the atoms under consideration, a and b, constants adjusted on diamond ($n_i = n_j = 3$), benzene, and graphite (in both $n = 2$). Actually, this kind of formula is subject to criticism, because in benzene and graphite the distance results at the same time from constant localized molecular orbitals involving sp^2 hybrids ($n = 2$) and from the effect of the delocalized π-system. Now, in a conjugated hydrocarbon, the interatomic distance is obtained to an excellent approximation by the linear law [7,62]

$$d \, (\overset{\circ}{A}) = 1.52 - 0.19\, \ell \qquad (55)$$

In an acetylenic hydrocarbon the law is [62]

$$d \, (\overset{\circ}{A}) = 1.50 - 0.15\, \ell \qquad (56)$$

In the absence of π-bonds, $\ell = 0$. The distances that correspond just to the localized molecular orbitals are thus

$$1.52 \text{ Å for } sp^2 - sp^2 \text{ and } 1.50 \text{ for } sp - sp \quad .$$

In diamond, the distance is 1.54, which corresponds to sp_3 hybridization. Therefore, the distance that corresponds to localized molecular orbitals involving hybrids of degrees n_i, n_j, is related by the linear relationship

$$r_{ij} \text{ (Å)} = 1.48 + 0.0005 \ (n_i + n_j) \qquad (57)$$

This linear dependence is certainly preferable to that of type (54).

At any rate, at present, only calculations made by formulas of the type (54) are available for diamond and silicon [87]. The results are clear-cut: The changes in the plane-to-plane spacing with respect to the perfect crystal appear to decrease rapidly toward the bulk; only the first three or four planes are really affected. The contraction obtained is 13% in diamond and 8% in silicon for the top plane-pair. These results suggest a deformation that is larger in diamond than in silicon, which seems to be in disagreement with experimental results. It would be interesting to repeat these calculations using formula (57). Unfortunately, the coefficients are not easy to obtain. We know only the distance in silicon (2.34 Å). Consequently we have

$$r_{ij} \text{ (Å)} = 2.34 + B(n_i + n_j - 6) \qquad (58)$$

where B is an unknown constant. To a first approximation, one can assume that the ratio $\left[r(n=3) \right] \Big/ \left[r(n=1) \right]$ is the same for carbon and silicon compounds. Consequently we obtain for silicon

$$r_{ij} \text{ (Å)} \sim 2.25 + 0.015 \ (n_i + n_j) \qquad (59)$$

In gallium arsenide, the problem is more complicated. The gallium atoms do not play the same role as those of arsenic: For instance, on the (110) face, which is the stable one for the crystal, the As atoms are linked to only three Ga atoms, as in ordinary molecules containing arsenicum (e.g., AsH_3). In the absence of any charge displacement, the dangling hybrid orbital would be doubly used. The electronic repulsion between the charge associated with the nonbonding hybrid and the charges associated with bonding orbitals reduces the $\overparen{GaAsGa}$ angles, which explains the tendency of arsenic atoms to rise above the average surface plane. On the other hand, the sp_3 dangling orbitals of the Ga atoms, which should not be used, would tend to acquire a pure-p character as in BF_3. Thus, the hybridization of Ga tends to be of the sp_2 type: The $\overparen{AsGaAs}$ angles decrease, the Ga atoms shift to the inside.

A Hückel method study of the adsorption of a cesium atom on the (110) face of GaAs [38], carried out in order to determine the adsorption energy of a cesium atom as well as the charge of that atom and the change in the ionization energy, assuming a surface deformation in agreement with the above scheme, shows that in order to obtain a positive net charge for the Cs atom and an ionization energy lower than that corresponding to a clean (110) face, the deformation must be very weak. The calculated adsorption energy is very close to the experimental energy: $\sim$ 2.5 eV. The As atoms are raised above the average surface plane by about 0.2 Å, and the Ga atoms move inward by about 0.2 Å. A direct study made by the maximum overlap criterion [88] gives values very close to those suggested by Hückel calculations. However, they remain slightly smaller than the LEED ones [82].

b) Metals

The system under study is now entirely delocalized. The preceding method no longer applies. However, the evaluation of bond orders permits a simple treatment of the problem. In fact, it is well known that bond orders are related to interatomic distances: An iterative calculation can thus give the geometry.

Baldock computed the bond orders for several kinds of lattices formed by alkaline-like atoms: planar square lattice, simple cubic lattice, bcc lattice [50]. He showed that the bond orders are much different in the neighborhood of the surface from those inside the lattice:

0.576   0.543   0.541   0.541   .....   0.540

0.329	0.461	0.436	0.434	.....	0.432
0.357	0.364	0.440	0.421	.....	0.417
...	...	...	...	...	...
0.360	0.389	0.397	0.400	.....	0.405

(the value 0.400 in the last line is corrected according to a recent calculation [89]).

However, Baldock did not draw conclusions from his important result. The sum of the bond orders associated with one line of the above diagram decreases as one moves away from the upper side. For the first five atoms that sum is, respectively, 2.20, 1.66, 1.58, ... and tends to 1.55. Assuming that the bond orders vary linearly with the equilibrium distance, these results mean that the crystal is distorted to a sort of cushion-shape, because the faces show a tendency to become round.

A study carried out on finite-sized clusters corresponding to the lattices studied by Baldock, based on the Hückel method and taking into account the changes of βs between neighboring atoms together with a bond-order distance relation of the type $r = a - b\ell$, let us (on convergency) to the following conclusions [90,91]:

An n×n planar square lattice has a tendency to round off its sides. The effect appears to be greater for a (1,0) side than for a (1,1) side. The distances between two parallel rows increase regularly from the boundary to the center. The results are thus in agreement with Baldock's and with the conclusions he could have drawn, as well as with the surface contractions observed for metals.

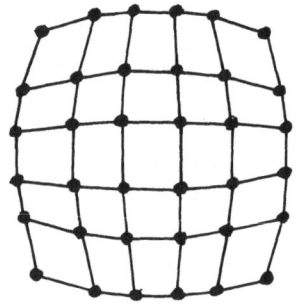

Geometry of square lattice: (1,0) sides.

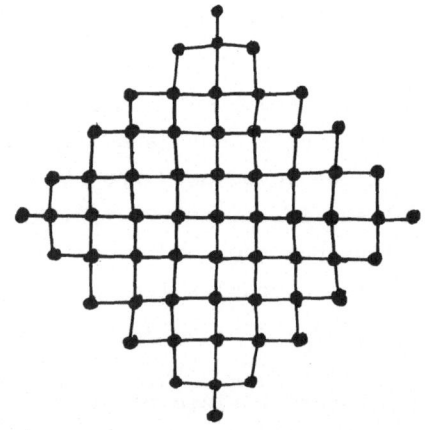

Geometry of square lattice: (1,1) sides.

These results also confirm the weak penetration of geometric relaxation, which does not affect more than a few rows of atoms. For the simple cubic lattice as well, the distortion phenomenon is quite evident.

Rectangular n×m clusters show a strong tendency to drop formation:

4 × 10

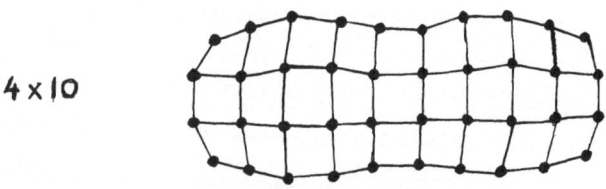

Calculations using a distance-bond order relationship of the homography type lead to the same conclusions [92].

The results just summarized clearly show that small clusters are far from corresponding to perfect lattices. The atoms located at the corners are very special. This explains the catalytic properties of metals in a highly divided state, the number of active sites (viz. corners) being greater, the greater the degree of division.

The presence of a surface produces in its neighborhood effects other than changes in interatomic distances. Such effects are of two types.

In transition metals, for instance, a marked loss of spherical symmetry in the electron density associated with $\underline{d}$ orbitals appears for atoms located on the surface. The effect is different depending on the face under consideration [51], because it is related to population differences in the various orbitals. A calculation made by the Hückel method on very small Ni clusters, in which the neighbors of the central atom are arranged according to the environment of an atom of the (111), (100), (110) surface, and taking into account the $\underline{4s}$ and $\underline{3d}$ orbitals, show that these differences are quite important. In fact, the small size of the clusters treated may be responsible for their importance, but the qualitative trend is certainly general.

Another consequence of the existence of faces is the change in the charges of atoms situated near them. In the classical Hückel method, the diagonal elements associated with the $\underline{s}$ orbital of an alkaline metal are the same for all the atoms (cf. Appendix A); in the SCF method, however, after orthogonalization of the atomic orbitals according to the Löwdin method, a weak but non-negligible contribution of the potential created by the neighboring neutral atoms must be added to the L_{pp} value given in Appendix A $\left\lfloor \text{Eq. (93)} \right\rfloor$. In fact, that formula presupposes that the core potential A^+, i.e., the potential of the atom without its valence electron, can be identified by the potential, changed in sign, of the electron density $\underline{a}^2$, which corresponds to the $\underline{s}$ orbital associated with it

$$(A^+, \sim - (a^2, \tag{60}$$

$(\rho,$ conventionally indicates the potential created by the electron repartition of density ρ; whereas, actually

$$(A, = (A^+, + (a^2, \tag{61}$$

The potentials created by the neutral atoms are very weak. In
practice, only the terms coming from the nearest neighbors bring a
non-negligible contribution. Therefore, the correction in question
brings about only a general shift of the α_r values for the bulk atoms,
and thus neither the molecular orbitals nor the charges and bond orders
are affected; however, the L_{pp} values of the surface atoms, which
have fewer neighbors, will be different from those of the bulk atoms.
In general, it can be shown that

$$\alpha_p = \alpha + \sum_r (R,p^2) \qquad\qquad (62)$$

R being the potential of the neighbor $\underline{r}$ of the atom $\underline{p}$ under
consideration. The integrals (R,p^2) — so-called <u>penetration integrals</u>—
are negative; the αs are negative; therefore, for a surface atom $\underline{s}$

$$\alpha_s > \alpha_i \qquad\qquad (63)$$

α_i being the value associated with the bulk atoms. The surface atoms
become less electronegative: They have a tendency to become poorer in
electrons.

This phenomenon is well known in molecules. SCF calculations
performed on the π-system of butadiene [93-94] lead to the following
π-charges:

$$q_1 = q_4 = 0.98 \quad\text{and}\quad q_2 = q_3 = 1.02$$

The terminal atoms (1 and 4) are electron-deficient.

A general modification in charge distribution is brought about
by the fact that the α integrals of the terminal atoms are modified
and are different from those of the others. Chemists have known for a
long time that the presence of a heteroatom (oxygen) at one end of a
conjugated chain —e.g., a polyene one— leads to the apparence of
alternating net charges on the atoms of the chain:

$$\begin{array}{ccccccc} - & + & - & + & - & + \\ O = C\!-\!C & = & C\!-\!C & = & C\!- \end{array}$$

In an unsubstituted polyene, the effect remains the same, but is
weaker; for instance, in $C_{16}H_{18}$, the net charges are [94]:

$$+0.020 \quad -0.018 \quad +0.002 \quad -0.003 \quad +0.001 \quad -0.001 \quad 0 \quad 0$$

In a linear chain of an alkaline metal, the effect will be the same [95]. Starting from one of the terminal atoms, an alternation of net charges will be observed, which is damped out toward the bulk. In solid-state physics this phenomenon is known as the Friedel oscillations. In the real three-dimensional metal the effect is likely to be more complex: In fact, the alternation of charges in a conjugated hydrocarbon molecule appears only if the molecule consists of chains and contains only even cycles [7,11]. This condition is not realized in the real bcc or cfc arrangements. It is difficult to predict a priori the charge distribution. The problem deserves further study.

c) Practically Ionic Lattices

In the classical theory, the displacement of the ions arises from polarization of the surface ions by the electrostatic field due to the termination of the bulk periodicity at the surface; as the polarizabilities of the two ionic species are different, their resulting displacements are different, and the surface becomes rumpled [96].

Independently of any calculation, it must be noted that the cellular model proposed by us for alkali halides [22] explains that the surface metal atoms move more than the halogen atom toward the interior of the crystal [79]. In fact, the metal atom has just five neighbors. Instead of retaining the sp_3d_2 hybridization, it will adopt the sp_3d hybridization, which demands a lower energy. The latter hybridization corresponds to a pyramidal arrangement, which demands that the metal atom should be below the four surface halogens to which it is bound.

In the case of a crystal where the anion is a polyatomic one (e.g., the carbonate ion), the surface effect does not reveal itself just by ion relaxation. We have seen that the crystal field produces slight modifications in the ions (carbonate in calcite); however, in the infinite crystal the general symmetry is preserved. On the other hand, for ions on the surface, the crystal field is different and has either a much lower symmetry or even no symmetry, as happens in the case of a cleavage face of calcite. A special kind of deformation must thus take place for these ions.

A preliminary calculation [27] on the surface carbonate ions of calcite, not taking into account either a possible change of the total charge (assumed to be equal to -2) or lattice relaxation, has shown that:

(1) the oxygen atom situated inside the crystal penetrates

further into it, the CO distance increasing by 4%, and

(2) the surface oxygen atom moves to the outside, thus
abandoning the surface,
so that the CO_3 group tends to break down, one oxygen moving into
the bulk, the other tending to form a CO_2 group. These deformations
explain the thermal decomposition of calcite, a phenomenon initiated
at the surface: CO_2 molecules come apart easily and the remaining
oxygens will continue to move inward and will settle between calcium
ions (CaO has the rock-salt structure).

3. Vacancy Problem

A problem connected with the existence of faces is the problem
of vacancies. A perfect crystal does not exist: Even aside from the
practical impossibility of preparing a crystal without impurities,
thermodynamics indicates that the presence of vacancies is inevitable
if the absolute temperature is not equal to zero [7]. The absence of an
element in the lattice destroys the periodicity and thus brings about
a modification of the geometry around the vacancy.

The question of determining that modification theoretically has
been tackled from a classical standpoint in the case of perfectly
ionic lattices (e.g., NaCℓ). The atoms located around the vacancy are
displaced toward the center of the vacancy. The shift is about 0.1 to
0.2 the lattice parameter [98,99,100].

An electron replacing a negative ion ($Cℓ^-$) in the lattice creates
an F center. Several theoretical studies have shown the importance of
the relaxation effect around the vacancy on F electronic levels.
However, the results obtained are contradictory, and it is difficult
at present to use them for obtaining a more detailed description of
the local geometry [101].

In the case of an alkali metal, the Hückel method used for
studying the effect of faces also applies to clusters with vacancies.
It shows two main effects: The vacancy tends to be resorbed, and, when
there are two vacancies, they tend to melt together [102]. The figure
below shows the importance of distortion with respect to a cluster of
the same size and without vacancy (cf. p.92).

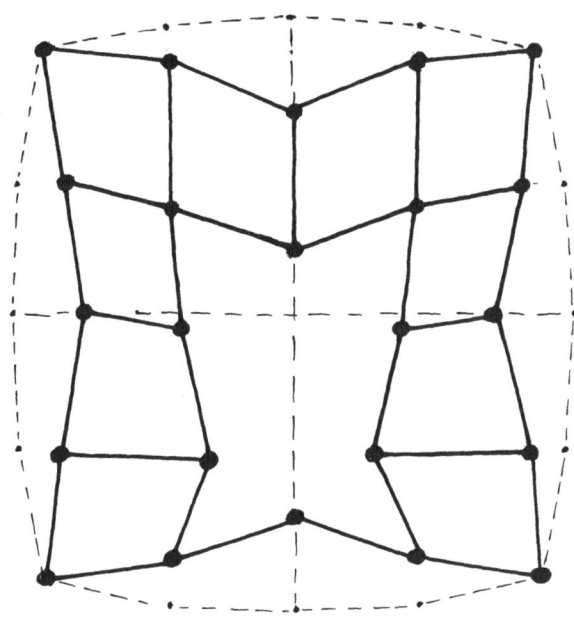

In the ion oxide FeO (which has the same structure as NaCℓ), the number of iron atoms is normally insufficient to fill all the corresponding sites, and the crystal is rich in vacancies. The vacancies probably occur in groups of four, occupying the vertices of a tetrahedron centered on an iron atom [23] that is sp_3-hybridized —contrary to the other lattice atoms, which are sp_3d_2-hybridized. The situation could be similar in NaCℓ.

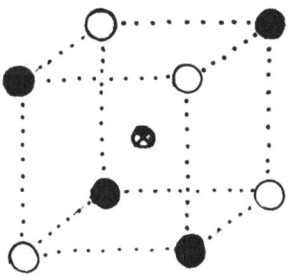

● oxygen atom

○ absent iron atom

⊗ iron atom, sp_3-hybridized

VI. External Shapes of Crystals

Undoubtedly one of the most fascinating features of crystals for the layman is the beauty and the variety of form, as well as their colors. Although the external aspect of a crystal is not independent of its structure, and is connected with the crystal system to which it belongs, the problem of the external shape is difficult to solve.

Classical thermodynamics allows a general approach to the problem. We shall not expound upon that theory in detail, for it is presented in classical textbooks [103]. Suffice it so say that it considers the energy as being the sum of two contributions, a volume energy, which depends on the number of constituting elements, and a surface energy, which is proportional to the number of surface elements.

In practice, the latter term appears to be proportional to the surface itself, the proportionality factor being related to be arrangement of the surface elements, viz. to the nature of the face under consideration. The theory of crystal shapes rests on the same principle as that of the surface tension of liquids. For a given number of elements, the form to be adopted will tend to minimize the surface energy. The minimization condition is simplified by the fact that the faces are planar: It reduces to the Wulff theorem [104].

$$\Sigma_{hk\ell} / d_{hk\ell} = Cte \qquad (64)$$

where $\Sigma_{hk\ell}$ is the surface energy per unit area of the given face $(hk\ell)$ and $d_{hk\ell}$ is the distance of the face under study to the center of the polyhedron formed by various faces.

The Wulff Eq. (64) also governs crystal growth.

The problem thus reduces to an evaluation of the surface constants Σ associated with the various crystal faces, for the various arrangements and for each substance.

1. Metals

The case of metals is particularly favorable for a theoretical treatment. Among the approaches that make the prediction of surface energies possible, the simplest one consists of studying the change in the cohesion energy of finite clusters of increasing sizes [70]. Let n denote the number of atoms: Then the total energy of the system can be written in the form

$$\varepsilon_n = n\,a + \text{correction terms} \qquad (65)$$

The difference between the exact value ε_n and the asymptotic value $\underline{n\,a}$ may be attributed to the existence in the finite structures of faces, edges, and corners. In fact, the ratio correction terms/$\underline{n\,a}$ tends to zero when $\underline{n}$ tends to infinity, and it is possible to represent it as sum of terms associated with the various geometric boundaries. For example, for a crystal whose faces are all of the same kind, Eq. (65) may be written

$$\varepsilon_n = n\,a + n_1\,\eta_1 + n_2\,\eta_2 + n_3\,\eta_3 \qquad (66)$$

where n_1 denotes the number of corner atoms, n_2 the number of edge atoms, n_3 the number of atoms situated on a face, but neither on a corner nor on a face ($a > 0$, $\eta_i < 0$).

The values of the increments η depend on the nature of the planes passing through the particular ion. To a first approximation, it is possible to write

$$\eta = \Sigma\, m_f\, b_f \qquad (67)$$

m_f indicating the number of f-type faces passing through this atom, and b_f indicating the increment relative to this kind of face. For an atom situated on a corner, such an approximation may be more questionable. However, for a given type of structure, the number of corners remains the same when $\underline{n}$ increases, so that the error soon vanishes. Using these assumptions, we have

$$\varepsilon_n = n\,a + \underset{i}{\Sigma}\,\underset{f}{\Sigma}\, m_f^i\, b_f \qquad (68)$$

in which m_f^i represents the number of f-type faces passing through a given atom $\underline{i}$. For instance, when all the faces are of the same kind, we have

$$\varepsilon_n = n\,a + n_1 b + n_2 \times 2b + n_3 \times 3b$$

$$= n\,a + (n_1 + 2n_2 + 3n_3)b \qquad (69)$$

(if the corners are placed at the intersection of three faces).

To determine the constants $\underline{a}$ and $\underline{b}$, it is necessary to study clusters having the same shape (cubes, octahedra, ...) with increasing n values. For a sufficiently large n, the cohesion energy may be represented by Eq. (69).

The application of the procedure just outlined to lithium clusters [70] leads to satisfactory values of the surface energy per atom b, in agreement with theoretical results obtained by more complicated methods and with experimental values for the liquid metal and the vapor pressure near the melting point (Tables 19 and 20). The values obtained for b satisfy the theorem of Wulff. For instance, for a cfc cuboctahedron bounded by (111) and (110) faces, one finds

$$d_{100}/d_{111} = 2\sqrt{3} = 1.15 \sim b_{100}/b_{111} = 1.2$$

Knowledge of the values of b makes it possible to know which will be the most stable faces, and therefore the shape the metal crystal will take up. For instance, for lithium the polyhedron formed by (110) faces only, will be the most stable.

Table 19

Lattice	Face	Surface energy (kcal/atom)	
		Calculated	Experimental values[105-106]
bcc	100	6.7	
	110	3.3	4.05 to 3.58
	111	6.3	
cfc	111	3.7	
	110	4.4	

Table 20
Theoretical values of surface energies of lithium
estimated by different authors

Authors	Surface energy (kcal/atom)	
Huang and Wyllie [107]	8.04	
Stratton [108]	3.6	
Skapki [109]	3.84	(110) bcc
Langard and Kohn [106]	3.42	(110) bcc
	3.24	(111) cfc
Baldock [50]	5.21	(100) bcc
	2.76	(110) cfc

In fact, in the method described here and applied to lithium, it is not necessary to go through surface energies in order to predict the shape of a cluster. It is sufficient to compute ε_n for several homothetic clusters of greater and greater size, so as to deduce by interpolation the most stable arrangement for a given number of atoms [70]. The computation of surface energies presented here is but a test of the method, for it permits comparison of the results with those of other methods.

We emphasize the following point. Calculations made on lithium by the Hückel method show that the energy per atom varies very little from one structure to the other (at most a few kcal/atom). Therefore, the transitions from one form to the other for small structures demand very small energies. These energies are of the order of the thermal energy. Consequently small clusters will behave more like liquid drops than like well-ordered rigid crystals. This explains the experiment which consists of observing the spontaneous transformation at 2500 K of the point of a tungsten needle into a crystalline face perpendicular to the direction of the needle [110]. The atoms situated at the end are sufficiently mobile for the rearrangement to take place.

Regardless of the external shape of the crystal and of the values of the surface energy, direct calculation of ε_ns for increasing sizes allows the derivation of the constant $\underline{a}$ $\left[\text{Eq. (65)}\right]$ relative to the various arrangements (bcc, cfc). When applied to lithium, the method does lead the preferential arrangement [70] (bcc). However, the curious fact appears that the bcc arrangement is only favored starting with a given size. For clusters of only a few hundred atoms, structures possessing fivefold symmetry axes (pentagonal pyramids, bipyramids...) are the most stable structures. Only beyond a critical size, probably of the order of 10^5-10^6 atoms —and thus still corresponding to very small structures, just visible under the electron microscope— does the crystal really deserve its name. Similarly, calculations made on nickel clusters [111] show that for a number of atoms lower than 150, icosahedra (symmetry of order 5) are more stable than cfc cuboctahedra possessing the same number of atoms. Although the number (150) may seem somewhat small, the phenomenon is the same.

Experimentally, quinary symmetry forms have actually been observed for small clusters of gold, silver, and iron [112]. These shapes increase up to a critical size (200 Å in diameter for gold), then suddenly coalesce to give crystal having the final symmetry.

2. Other Crystals

The theory of surface energy does not apply to crystals with localizable orbitals or to practically ionic systems.

For the latter, interesting results concerning the external shape can be obtained by elementary classical arguments, which we shall not treat here (cf., e.g., [113]).

We shall rather indicate a result [114] that is interesting because of its simplicity. Given an infinite perfectly ionic crystal, let us trace around a given ion a simply connected surface $\underline{S}$. The electrostatic potential to which the given ion is subjected may be broken down into two terms: the potential created by the ions situated within the surface $\underline{S}$ and that created by charges outside $\underline{S}$

$$V = V_i + V_e \qquad (70)$$

A detailed study of the potential V_e for arrangements obtained by continuous deformation from the cfc lattice to the bcc one, passing through the rhombohedral arrangement, shows that, for a given number of ions inside the surface $\underline{S}$, the external-charges potential V_e reaches a minimum when $\underline{S}$ corresponds to the natural-cleavage polyhedron: a cube for the cfc and bcc lattices, a rhombohedron for intermediate forms.

To obtain a finite crystal from an infinite one, work must be done against the external potential that corresponds to the desired polyhedron. Because that potential is at a minimum for a cube in NaCℓ or in CsCℓ, and for a given rhombohedron in calcite, it is clear why these crystals should adopt these cleavage shapes. Note that in calcite the rhombohedron corresponding to the surface $\underline{S}$ is not the rhombohedron that generates the lattice in crystallography [66], but the rhombohedron of usual natural crystals.

For crystals with localizable molecular orbitals, such as diamond, the problem is simpler. Cleavage will take place along planes that will leave the greatest number of bond orbitals unchanged, so as to modify the energy as little as possible. Of course, complications arise from geometric relaxation. For diamond, where that phenomenon does not take place, (111) cleavage is practically the only one. In contrast, the (110) face is more stable than the (111) face in gallium arsenide.

In corundum the (0001) plane (horizontal plane, Fig.4, p.33), which contains only oxygen atoms, is a cleavage plane which is very easy to obtain. Each surface oxygen atom is linked to two internal aluminum atoms and carries two nonbonding orbitals. That is the usual oxygen structure (cf. H_2O).

3. Twin crystals

Independently of crystal tangle due to accidental conditions of growth, many crystals seem to be constituted by joining of two crystals, more or less extended, symmetrical with respect to a plane. In fact, the sample is a sole crystal —giant molecule— which possesses translation symmetries only in each part, not in the whole. Such crystals are called twin-crystals. The symmetry plane is the twin-plane. This plane corresponds necessarily to a reticular plane of each part. Nevertheless, the joining of the lattices of the two parts, which must be obtained without discontinuity in the molecular orbital system, provokes in general distortions in the neighborhood of the twin-plane.

For example, in zinc-blende a strict symmetry with respect to a (111) plane (base plane, Fig.3, p.32) would place the neighboring sulfur atoms too close to each other, and zinc atoms linked to six sulfur atoms. This situation is impossible. In reality, the sulfur atoms come into the zinc atom plane. The zinc atoms are sp_2-hybridized and the sulfur atoms are sp_3d -hybridized (cf. PF_5). The second part of the crystal can be built without difficulty from this base plane, symmetrical with respect to the first crystal part.

In rutile, the twin-plane is a (101) plane (Fig.4, p.33). As in ZnS, the oxygen atoms located in the neighborhood of the twin-plane, come into this plane. The titanium atoms corresponding to the atoms located at the cell center, possess only four neighbors, located at corners of a weakly distorted tetrahedron. They are sp_3-hybridized (cf. $TiCl_4$).

Conclusions and Various Outlooks

The examples evoked in this lecture demonstrate the advantages that can be derived from the concepts and methods of molecular quantum chemistry in order to explain crystal structure and to predict the shape of the crystals. Other possibilities, equally interesting, are offered by this theory. The above study is constructed on the prediction of the molecular orbitals that are used in the ground state. These orbitals are responsible for the cohesion of the system. They indicate the orbitals that do not participate (or only minimally) in the formation of the lattice and the orbitals that are not used. Knowledge of these orbitals permits one to obtain information on the electronic and magnetic properties. Our purpose is not to study this important set of properties. We shall only give same simple examples to show why the general description which we have given, can be used in these problems.

When the number of electrons corresponds exactly to the number of bonding molecular orbitals, the atomic orbitals that are not used to build the hybrid or anti-bonding molecular orbitals corresponding to bond orbitals, remain unused in the ground state. The crystal is an insulator (e.g., diamond, Al_2O_3, TiO_2, CaF_2, Be_2SiO_4, $CaCO_3$...). The crystal will become a conductor as a result of radiations sufficiently high in energy to provoke a jump into an excited state.

The problem becomes interesting when the number of electrons is greater than in the above case. For example, in ReO_3, three orbitals, d_{xy}, d_{yz}, d_{zx}, remain on each metal atom. These orbitals, owing to their symmetry, combine with the nonhybridized 2$\underline{p}$ orbitals of the oxygen atoms to build a completely delocalized molecular orbital system.

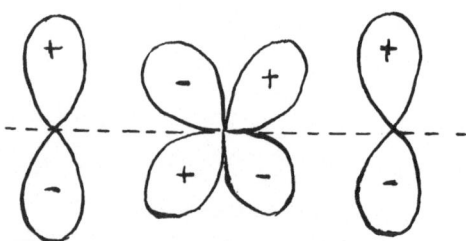

The energy associated with 2$\underline{p}$ oxygen orbitals is much lower than that associated with a $\underline{d}$-metal orbital. Consequently, the two sets of orbitals —the 2$\underline{p}$ orbitals of the oxygens and the $\underline{d}$ orbitals of the

metals— are not mixed. In practice, one can consider that the 2p
orbitals of each oxygen are doubly used nonbonding localized orbitals
and that the delocalized system is constructed only by the d orbitals.
These orbitals form a band located over the doubly used bonding
orbitals.

In Re_2O_3, each rhenium atom contributes one electron to this
band. That is 1/6 full. Re_2O_3 will have a metallic character. The
first excited electron levels are located immediately over the ground
state. They form a continuous set. ReO_3 is black.

The tungsten atom possesses one less electron. In WO_3, which has
the same structure as ReO_3, the d band is empty. WO_3 is a semi-conductor.
The first excited levels are located at a specific distance from the
highest used levels in the ground state: WO_3 is yellow.

In CuO, the sp_2d hybridization leaves five atomic orbitals unused.
Nine electrons per copper atom are available: CuO is a black semi-
-conductor.

The case of crystals we have described by means of a cellular
model is similar. NaCl and MgO are insulators because all the electrons
are used in bonding orbitals of cells. In contrast, in VO, each
vanadium atom possesses three electrons, which use completely
delocalized molecular orbitals constructed from the d_{xy}, d_{yz}, and d_{zx}
orbitals of the vanadium atom. VO is a conductor.

Similar reasoning would result in an explanation of the magnetic
or ferromagnetic properties.

x

x   x

In 1824, Beudant wrote, in his Traité de Minéralogie: "Si les
découvertes de la cristallographie ont fait sortir la minéralogie de
son empirisme, ..., les progrès de la Chimie et de la Physique lui
ont rendu des services non moins importants, et l'ont réellement
élevée au rang des Sciences exactes". One and a half centuries later,
this affirmation is still with us. In addition, the proposed analysis
is interesting because it unifies the treatment of two states of
matter: molecules and crystals. Usually these two areas are considered
separately, though it is obvious that the laws which control the
structure of matter are the same for all structures.

Appendix A

The self-consistent field method and the Hückel approximation

We have included only some general results useful for the comprehension of the text. More details are given in references [7] and [11].

1) Expression of total electron energy as function of φ_i molecular orbitals doubly used in the ground state:

$$E_e = 2 \sum_i I_i + \sum_i \sum_j (2J_{ij} - K_{ij}) \tag{71}$$

where $I_i = \langle \varphi_i \mid T+U \mid \varphi_i \rangle$

T = kinetic energy operator

U = operator corresponding to the potential created by the nuclei

$J_{ij} = \langle \varphi_i^2 \mid \frac{1}{r} \mid \varphi_j^2 \rangle$ (Coulomb integral)

$K_{ij} = \langle \varphi_i \varphi_j \mid \frac{1}{r} \mid \varphi_i \varphi_j \rangle$ (exchange integral)

2) Expression of E_e as a function of atomic orbitals:

$$\varphi_i = \sum_r c_{ir} \chi_r \qquad (r = 1 \text{ to } n) \tag{5}$$

$$E_e = 2\sum_i \sum_p \sum_q c_{ip} c_{iq} I_{pq} + \sum_i \sum_j \sum_p \sum_q \sum_r \sum_s c_{ip} c_{iq} c_{jr} c_{js} \Big[2\,(pq,rs) - (ps,rq) \Big] \tag{72}$$

where $I_{pq} = \langle \chi_p \mid T+U \mid \chi_q \rangle$

$(pq,rs) = \langle \chi_p \chi_q \mid \frac{1}{r} \mid \chi_r \chi_s \rangle$

3) Minimization of E_e with respect to the $\underline{c}$ coefficients gives the homogeneous linear system:

$$\begin{cases} \cdots \cdots \cdots \\ \sum_p c_{ip} (L_{pq} - eS_{pq}) = 0 \qquad (q = 1 \text{ to } n) \\ \cdots \cdots \cdots \end{cases} \tag{73}$$

where $S_{pq} = \langle \chi_p \chi_q \rangle$ and $L_{pq} = I_{pq} + G_{pq}$

with $G_{pq} = \underset{j}{\Sigma} \underset{r}{\Sigma} \underset{s}{\Sigma} c_{jr} c_{js} \left[2(pq,rs) - (ps,rq) \right]$ (74)

The compatibility condition of system (73) —so-called <u>secular equation</u>

$$\det |L_{pq} - e S_{pq}| = 0 \qquad (75)$$

gives $\underline{n}$ roots, e_i. A molecular orbital φ_i corresponds to each root.

4) Another expression of electron energy:

$$E_e = \underset{i}{\Sigma} (I_i + e_i) \qquad (76)$$

(i = doubly used molecular orbital)

5) Orthogonalization of the atomic orbitals basis:

In general, the orbitals χ_r are not orthogonal. In the calculations it is interesting to use orthogonal orbitals. The most interesting procedure for obtaining orthogonal orbitals is the Landshoff-Löwdin procedure, which respects to a greater degree the initial orbitals:

$$(\chi') = (\mathbb{1} + \mathcal{S})^{-1/2} (\chi) \qquad (77)$$

being the matrix: $(S_{pq} - \delta_{pq})$ (δ = Kronecker's symbol).

Neglecting the S^2 terms, we have:

$$\begin{cases} \chi'_p = \chi_p - \dfrac{1}{2} \underset{k \neq p}{\Sigma} S_{kp} \chi_k \\[2mm] \chi'^2_p = \chi^2_p \end{cases} \qquad (78)$$

6) The Mulliken approximation

The potential created by the $\chi_p \chi_q$ cross-distribution can be replaced by the one created by

$$\frac{1}{2} (\chi^2_p + \chi^2_q) S_{pq} . \qquad (79)$$

This approximation is valid when the χ_p and χ_q orbitals are of the same nature and are appropriately located: $\underline{s}$ or π orbitals, hybrid orbitals pointing toward each other in twos.

Formulas (78) and (79) allow the simplification of the expression of secular equation elements corresponding to orthogonal orbitals:

$$
\left\{
\begin{array}{l}
S_{pq} \longrightarrow \delta_{pq} \\[2mm]
L_{pp} \quad \text{practically depends only on the nature of } \underline{p}\text{-orbital,} \\[2mm]
L_{pq} \quad \text{depends only on the nature of orbitals } \underline{p} \text{ and } \underline{q} \text{ and on} \\
\qquad \text{their distance.}
\end{array}
\right.
$$

The elements L of Eq. (73) corresponding to this orthogonalized basis are transferable from one molecule to another. Herein lies the interesting aspect of orthogonalization.

7) Hückel method

As early as 1931, Hückel wrote a secular equation similar to Eq. (73). Not knowing the values of the various elements L, without theoretical justification, Hückel hypothesized that the overlaps can be neglected (i.e., $S_{pq} = \delta_{pq}$), that the diagonal terms ($L_{pp} = \alpha_p$) depend only on the nature of the corresponding atom P, and that the nondiagonal elements ($L_{pq} = \beta_{pq}$) depend only on the nature of P and Q atoms and on the distance PQ. In addition, the β_{pq}s corresponding to nonneighboring atoms are neglected. These hypotheses are in agreement with SCF results. The last approximation is justified by the rapid decrease of L_{pq} terms corresponding to orthogonalized orbitals.

The β_{pq}s corresponding to orthogonalized orbitals (χ') are approximately proportional to the overlap integrals S_{pq} corresponding to initial orbitals (χ) before the orthogonalization[7]:

$$
\beta_{pq} \sim S_{pq} \tag{80}
$$

The Hückel method gives molecular orbitals very close to the SCF molecular orbitals, but the energy is different. According to Hückel, the electron energy is

$$
E_e^H = 2 \sum_i e_i \tag{81}
$$

In reality, the expression of this energy is different (76). From Eqs. (75) and (74) we obtain:

$$E_e = 2 \sum_i e_i - 2 \sum_i \sum_p \sum_q c_{ip} c_{iq} G_{pq} \tag{82}$$

The electron charges given by Eq. (15), owing to the orthogonalization, become:

$$q_r = 2 \sum_i c_{ir}^2 \tag{83}$$

In addition, we shall define the bond order:

$$\ell_{rs} = 2 \sum_i c_{ir} c_{is} \tag{84}$$

corresponding to orthogonalized orbitals. Consequently in this orthogonalized basis, we have:

$$\begin{cases} G_{pp} = \frac{1}{2} q_p J_{pp} + \sum_{k \neq p} q_k J_{pk} \\[2mm] G_{pq} = -\frac{1}{2} \ell_{pq} J_{pq} \end{cases} \tag{85}$$

On the other hand, the nuclear repulsion is:

$$E_n = \sum_{(pq)} \frac{n_p n_q}{r_{pq}} \sim \sum_{(pq)} n_p n_q J_{pq} \tag{86}$$

(n_p = nuclear charge of atom P)

The total energy ($E_e + E_n$) is

$$E = 2 \sum_i e_i - \frac{1}{4} \sum_p q_p^2 J_{pp} + \sum_{(pq)} (n_p n_q - q_p q_q) J_{pq} + \frac{1}{2} \sum_{(pq)} \ell_{pq}^2 J_{pq} \tag{87}$$

If the net charges are equal to zero or very weak ($q_p \sim n_p$), we obtain:

$$E = 2 \sum_i e_i - \frac{1}{4} \sum_p n_p^2 J_{pp} + \frac{1}{2} \sum_{(pq)} \ell_{pq}^2 J_{pq} \tag{88}$$

In the Hückel method, the electron energy was

$$E_e = 2 \sum_i e_i = \sum_p q_p \alpha_p + 2 \sum_{(pq)} \ell_{pq} \beta_{pq} \tag{89}$$

If $q_n \backsim n_p$, the total energy is

$$E = \sum_p n_p (\alpha_p - \frac{n_p}{4} J_{pp}) + 2 \sum_{(pq)} \ell_{pq} \beta_{pq} + \frac{1}{2} \sum_{(pq)} \ell_{pq}{}^2 J_{pq} \tag{90}$$

If all the atoms are of the same nature

$$E = N(\alpha - \frac{n^2}{4} J_{pp}) + 2 \sum_{(pq)} \ell_{pq} \beta_{pq} + \frac{1}{2} \ell_{pq}{}^2 J_{pq} \tag{91}$$

where $\underline{N}$ is the number of atoms.

In the case where $n = 1$ (alkaline metals), Eq. (91) becomes

$$E = N(\alpha - \frac{1}{4} J_{pp}) + 2 \sum_{pq} \ell_{pq} \beta_{pq} + \frac{1}{2} \ell_{pq}{}^2 J_{pq} \tag{20}$$

8) Choice of Hückel's parameters:

The practical utilization of the Hückel method depends on the values of parameters α and β.

In the case where each orbital contributes only one electron, the L_{pp} terms can be written

$$L_{pp} \backsim - (\mathcal{J}_p + \frac{1}{2} J_{pp}) - \frac{1}{2} Q_p J_{pp} - \sum_{k \neq p} Q_k J_{pk} \tag{92}$$

(Q_p = net charge , $\mathcal{J}_p$ = ionization energy of atom P).
If the net charges are equal to zero or very weak, we obtain

$$L_{pp} \backsim - \mathcal{J}_p + \frac{1}{2} J_{pp} \tag{93}$$

To a first approximation [68], the integral J_{pp} can be written [115]

$$J \backsim \mathcal{J} - \mathcal{A} \tag{94}$$

Consequently

$$\alpha_p = L_{pp} \backsim - \frac{1}{2} (\mathcal{J} + \mathcal{A})_p \tag{95}$$

$|\alpha_p|$ appears as equal to the Mulliken electronegativity [7]. We note the difference from the erroneous but generally used expression $\alpha \backsim - \mathcal{J}$.

$\mathcal{J}$ and $\mathcal{A}$ are obtained from atomic spectroscopy [32]. Consequently the parameters α corresponding to neutral atoms are directly accessible.

If the χ_p orbital brings two electrons [11], we have

$$\alpha_p \sim - 2 \mathcal{J}_p \tag{96}$$

For the βs, **one generally** uses empirical relations. For example

$$\beta_{pq} = k \, \frac{\alpha_p + \alpha_q}{2} \, S_{pq} \tag{97}$$

where <u>k</u> is a constant of about 1.5 to 2.

9) Application to crystal theory

In an alkaline metal (or Cu, Ag, Au), all the parameters α are equal. The roots of the secular equation (75) form a continuous set and are symmetrically located with respect to the value α.

The number of the levels is equal to the number of the atoms. Each atom brings one electron. Consequently, in the ground state, the half of the levels is only utilized by the electrons. The Fermi level corresponds to the energy $e_F = -\alpha$. The work function is $\Phi = -\alpha$. We have seen that $|\alpha|$ is equal to the Mulliken electronegativity <u>x</u> of the isolated atom. Consequently we obtain the remarkable equality:

$$\Phi \sim x \tag{98}$$

This relation has been recently indicated [116], without a theoretical justification.

In a semi-conductor (silicon), the electron wave function is obtained from localized molecular orbitals. These orbitals are built from hybridized atomic orbitals (sp_3 in silicon). If we neglect the interactions between the orbitals which do not point at one another. we obtain only two values for <u>e</u>. These values are infinitely degenerate. The interactions between the orbitals give a spliting of the degenerate levels in two <u>bands</u>.

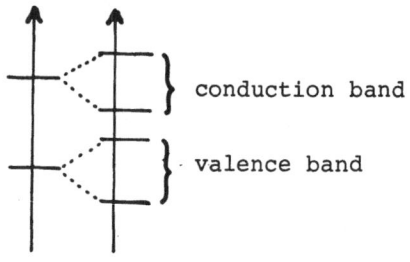

The orbitals of the inferior band are all doubly utilized in the ground state. They form the <u>valence</u> band. The orbitals of the other band remain unutilized. They form the <u>conduction band</u>.

The highest level of the valence band corresponds to an energy α + B, and the lowest of the conduction band to an energy α - B. Consequently, the energy of the Fermi level is equal to α. We have always: Φ = x. However, it is fundamental to remark that the electronegativity is the one of the sp_3 hybridized atomic orbital and not the one of the isolated atom. This distinction explains the difference between the actual work function Φ and the atom electronegativity [117].

Expressions of hybrid atomic orbitals

Hybridization s×p

1) sp$_3$ (symmetric):

$$t_1 = \frac{1}{2}(s + p_x + p_y + p_z)$$

$$t_2 = \frac{1}{2}(s + p_x - p_y - p_z)$$

$$t_3 = \frac{1}{2}(s - p_x + p_y - p_z)$$

$$t_4 = \frac{1}{2}(s - p_x - p_y + p_z)$$

$$\widehat{(t_i, t_j)} = 109,5°$$

$$t_1 = t_2 = t_3 = t_4$$

2) sp$_2$ (symmetric):

$$t_1 = \frac{1}{\sqrt{3}}s + \sqrt{\frac{2}{3}}p_x$$

$$t_2 = \frac{1}{\sqrt{3}}s - \frac{1}{\sqrt{6}}p_x + \frac{1}{\sqrt{2}}p_y$$

$$t_3 = \frac{1}{\sqrt{3}}s - \frac{1}{\sqrt{6}}p_x - \frac{1}{\sqrt{2}}p_y$$

$$t_4 = p_z$$

$$\widehat{(t_1, t_2)} = \widehat{(t_2, t_3)} = \widehat{(t_3, t_4)}$$
$$= 120°$$

$$t_4 \perp t_1, t_2, t_3$$

$$t_1 = t_2 = t_3 \neq t_4$$

3) sp (symmetric):

$$t_1 = \frac{1}{\sqrt{2}}(s + p_x)$$

$$t_2 = \frac{1}{\sqrt{2}}(s - p_x)$$

$$t_3 = p_y$$

$$t_4 = p_z$$

$$t_3 \perp t_1, t_2$$

$$t_3 \perp t_4$$

$$t_1 = t_2$$

Hybridization s×p×d

ν = number of vertices of the coordination polyhedron

1) $\nu = 4$ (square): $sp_2d_{x^2-y^2}$

$$
\begin{cases}
t_1 = \frac{1}{2} s + \frac{1}{\sqrt{2}} p_x + \frac{1}{2} d_{x^2-y^2} \\[2mm]
t_2 = \frac{1}{2} s - \frac{1}{\sqrt{2}} p_x + \frac{1}{2} d_{x^2-y^2} \\[2mm]
t_3 = \frac{1}{2} s + \frac{1}{\sqrt{2}} p_x - \frac{1}{2} d_{x^2-y^2} \\[2mm]
t_4 = \frac{1}{2} s - \frac{1}{\sqrt{2}} p_x - \frac{1}{2} d_{x^2-y^2}
\end{cases}
$$

$$(t_1 = t_2 = t_3 = t_4)$$

2) $\nu = 5$

a) $sp_3d_{z^2}$ (trigonal bipyramid) [9]

$$
\begin{cases}
t_1 = \sqrt{\frac{1-2a^2}{3}}\, s + \sqrt{\frac{2}{3}}\, p_y - a\sqrt{\frac{2}{3}}\, d_{z^2} \\[3mm]
t_2 = \sqrt{\frac{1-2a^2}{3}}\, s + \frac{1}{\sqrt{2}}\, p_x - \frac{1}{\sqrt{6}}\, p_y - a\sqrt{\frac{2}{3}}\, d_{z^2} \\[3mm]
t_3 = \sqrt{\frac{1-2a^2}{3}}\, s - \frac{1}{\sqrt{2}}\, p_x - \frac{1}{\sqrt{6}}\, p_y - a\sqrt{\frac{2}{3}}\, d_{z^2} \\[3mm]
t_4 = as + \frac{1}{\sqrt{2}}\, p_z + \sqrt{\frac{1}{2} - a^2}\, d_{z^2} \\[3mm]
t_5 = as - \frac{1}{\sqrt{2}}\, p_z + \sqrt{\frac{1}{2} - a^2}\, d_{z^2}
\end{cases}
$$

$$(a < 1/\sqrt{2}) \quad t_1 = t_2 = t_3 \neq t_4 = t_5$$

$$\widehat{(t_1,t_2)} = \widehat{(t_2,t_3)} = \widehat{(t_3,t_4)} = 120°$$

$$\widehat{(t_4,t_5)} = 180° \perp t_1, t_2, t_3$$

b) sp_3d_{xy} (or $sp_3d_{x^2-y^2}$) (square pyramid)

$$
\begin{cases}
t_1 = as + \frac{1}{2}(p_x + p_y) - \sqrt{\frac{1}{4} - a^2}\, p_z + \frac{1}{2} d_{xy} \\[2mm]
t_2 = as + \frac{1}{2}(-p_x + p_y) - \sqrt{\frac{1}{4} - a^2}\, p_z - \frac{1}{2} d_{xy} \\[2mm]
t_3 = as + \frac{1}{2}(-p_x - p_y) - \sqrt{\frac{1}{4} - a^2}\, p_z + \frac{1}{2} d_{xy} \\[2mm]
t_4 = as + \frac{1}{2}(p_x - p_y) - \sqrt{\frac{1}{4} - a^2}\, p_z - \frac{1}{2} d_{xy} \\[2mm]
t_5 = \sqrt{1 - 4a^2}\, s + 2\, a\, p_z
\end{cases}
$$

$$(a < \tfrac{1}{2}) \quad t_1 = t_2 = t_3 = t_4 \neq t_5$$

3) $\nu = 6$

a) sp_3d_2 (quadratic symmetry) [9]

$$
\begin{cases}
t_1 = as \qquad\qquad + \frac{1}{\sqrt{2}} p_z + \sqrt{\frac{1}{2} - a^2}\, d_{z^2} \\[2mm]
t_2 = as \qquad\qquad - \frac{1}{\sqrt{2}} p_z + \sqrt{\frac{1}{2} - a^2}\, d_{z^2} \\[2mm]
t_3 = \frac{1}{2}\sqrt{1 - 2a^2}\, s + \frac{1}{\sqrt{2}} p_x + \frac{1}{2} d_{x^2-y^2} - \frac{a}{\sqrt{2}} d_{z^2} \\[2mm]
t_4 = \frac{1}{2}\sqrt{1 - 2a^2}\, s + \frac{1}{\sqrt{2}} p_y - \frac{1}{2} d_{x^2-y^2} - \frac{a}{\sqrt{2}} d_{z^2} \\[2mm]
t_5 = \frac{1}{2}\sqrt{1 - 2a^2}\, s - \frac{1}{\sqrt{2}} p_x + \frac{1}{2} d_{x^2-y^2} - \frac{a}{\sqrt{2}} d_{z^2} \\[2mm]
t_6 = \frac{1}{2}\sqrt{1 - 2a^2}\, s + \frac{1}{\sqrt{2}} p_y - \frac{1}{2} d_{x^2-y^2} - \frac{a}{\sqrt{2}} d_{z^2}
\end{cases}
$$

$$(a < 1/\sqrt{2}) \quad t_1 = t_2 \neq t_3 = t_4 = t_5 = t_6$$

b) sp_3d_2 (octahedric symmetry)[9]

$$a = 1/\sqrt{6} \quad , \quad t_1 = t_2 = \ldots = t_6$$

c) sp_3d_2 (trigonal prism)

$$\begin{cases} t_1 = \dfrac{1}{\sqrt{6}}\ (s + p_z) + \dfrac{1}{\sqrt{3}}\ p_x \qquad\qquad + \dfrac{1}{\sqrt{2}}\ d_{zx} \\[2mm] t_2 = \dfrac{1}{\sqrt{6}}\ (s + p_z) - \dfrac{1}{2\sqrt{3}}\ p_x + \dfrac{1}{2}\ p_y + \dfrac{1}{2}\ d_{yz} \\[2mm] t_3 = \dfrac{1}{\sqrt{6}}\ (s + p_z) - \dfrac{1}{2\sqrt{3}}\ p_x - \dfrac{1}{2}\ p_y - \dfrac{1}{2}\ d_{yz} \\[2mm] t_4 = \dfrac{1}{\sqrt{6}}\ (s - p_z) + \dfrac{1}{\sqrt{3}}\ p_x \qquad\qquad - \dfrac{1}{\sqrt{2}}\ d_{zx} \\[2mm] t_5 = \dfrac{1}{\sqrt{6}}\ (s - p_z) - \dfrac{1}{2\sqrt{3}}\ p_x + \dfrac{1}{2}\ p_y - \dfrac{1}{2}\ d_{yz} \\[2mm] t_6 = \dfrac{1}{\sqrt{6}}\ (s - p_z) - \dfrac{1}{2\sqrt{3}}\ p_x - \dfrac{1}{2}\ p_y + \dfrac{1}{2}\ d_{yz} \end{cases}$$

$$(t_1 = t_2 = t_3 = t_4 = t_5 = t_6)$$

d) spd_4 (trigonal prism)

$$\begin{cases} t_1 = \dfrac{1}{\sqrt{6}}\ (s + p_z) \qquad\qquad\qquad + \dfrac{1}{\sqrt{3}}\ (d_{zx} + d_{x^2-y^2}) \\[2mm] t_2 = \dfrac{1}{\sqrt{6}}\ (s + p_z) + \dfrac{1}{2}\ (-d_{xy} + d_{yz}) - \dfrac{1}{2\sqrt{3}}\ (d_{zx} + d_{x^2-y^2}) \\[2mm] t_3 = \dfrac{1}{\sqrt{6}}\ (s + p_z) + \dfrac{1}{2}\ (d_{xy} - d_{yz}) - \dfrac{1}{2\sqrt{3}}\ (d_{zx} + d_{x^2-y^2}) \\[2mm] t_4 = \dfrac{1}{\sqrt{6}}\ (s - p_z) \qquad\qquad\qquad - \dfrac{1}{\sqrt{3}}\ (d_{zx} - d_{x^2-y^2}) \\[2mm] t_5 = \dfrac{1}{\sqrt{6}}\ (s - p_z) + \dfrac{1}{2}\ (-d_{xy} - d_{yz}) + \dfrac{1}{2\sqrt{3}}\ (d_{zx} - d_{x^2-y^2}) \\[2mm] t_6 = \dfrac{1}{\sqrt{6}}\ (s - p_z) + \dfrac{1}{2}\ (d_{xy} + d_{yz}) + \dfrac{1}{2\sqrt{3}}\ (d_{zx} - d_{x^2-y^2}) \end{cases}$$

(This hybridization is less probable than the sp_3d_2
hybridization because it requires higher energy).

4) $\nu = 7$ (e.g., NbF_7^{2-})

$$t_1 = as + dp_z + \sqrt{1-a^2-d^2}\, d_{z^2}$$

$$t_2 = bs + \frac{1}{2} p_x + cp_y + ep_z + \frac{1}{2} d_{xy} + \sqrt{\frac{1}{4} - c^2}\, d_{yz} - \sqrt{\frac{1}{4} - b^2 - e^2}\, d_{z^2}$$

$$t_3 = bs - \frac{1}{2} p_x + cp_y + ep_z - \frac{1}{2} d_{xy} + \sqrt{\frac{1}{4} - c^2}\, d_{yz} - \sqrt{\frac{1}{4} - b^2 - e^2}\, d_{z^2}$$

$$t_4 = bs - \frac{1}{2} p_x - cp_y + ep_z + \frac{1}{2} d_{xy} - \sqrt{\frac{1}{4} - c^2}\, d_{yz} - \sqrt{\frac{1}{4} - b^2 - e^2}\, d_{z^2}$$

$$t_5 = bs + \frac{1}{2} p_x - cp_y + ep_z - \frac{1}{2} d_{xy} - \sqrt{\frac{1}{4} - c^2}\, d_{yz} - \sqrt{\frac{1}{4} - b^2 - e^2}\, d_{z^2}$$

$$t_6 = \sqrt{\frac{1-a^2-4b^2}{2}}\, s + \sqrt{\frac{1}{2} - 2c^2}\, p_y - \sqrt{\frac{1-d^2-4e^2}{2}}\, p_z - c\sqrt{2}d_{yz}$$
$$+ \sqrt{\frac{a^2+4b^2+d^2+4e^2-1}{2}}\, d_{z^2}$$

$$t_7 = \sqrt{\frac{1-a^2-4b^2}{2}}\, s - \sqrt{\frac{1}{2} - 2c^2}\, p_y - \sqrt{\frac{1-d^2-4e^2}{2}}\, p_z + c\sqrt{2}d_{yz}$$
$$+ \sqrt{\frac{a^2+4b^2+d^2+4e^2-1}{2}}\, d_{z^2}$$

$c = 0$ or $1/2$. The $\underline{a}$, $\underline{b}$, $\underline{d}$, $\underline{e}$ coefficients are connected by the three relations $\langle t_1 t_2 \rangle = \langle t_1 t_6 \rangle = \langle t_2 t_6 \rangle = 0$.

5) $\nu = 8$

    a) sp_3d_4 (dodecahedron)

$$t_1 = \frac{1}{\sqrt{8}} s - ap_y + \sqrt{\frac{3}{8} - a^2}\, p_z - \sqrt{\frac{1}{2} - a^2}\, d_{yz} - \sqrt{a^2 - \frac{1}{8}}d_{x^2-y^2} + \frac{1}{\sqrt{8}}d_{z^2}$$

$$t_2 = \frac{1}{\sqrt{8}} s + ap_y + \sqrt{\frac{3}{8} - a^2}\, p_z + \sqrt{\frac{1}{2} - a^2}\, d_{yz} - \sqrt{a^2 - \frac{1}{8}}d_{x^2-y^2} + \frac{1}{\sqrt{8}}d_{z^2}$$

$$t_3 = \frac{1}{\sqrt{8}} s + ap_x - \sqrt{\frac{3}{8} - a^2}\, p_z - \sqrt{\frac{1}{2} - a^2}\, d_{zx} + \sqrt{a^2 - \frac{1}{8}}d_{x^2-y^2} + \frac{1}{\sqrt{8}}d_{z^2}$$

$$t_4 = \frac{1}{\sqrt{8}} s - ap_x - \sqrt{\frac{3}{8} - a^2}\, p_z + \sqrt{\frac{1}{2} - a^2}\, d_{zx} + \sqrt{a^2 - \frac{1}{8}}d_{x^2-y^2} + \frac{1}{\sqrt{8}}d_{z^2}$$

$$t_5 = \frac{1}{\sqrt{8}} s + \sqrt{\frac{1}{2} - a^2}\, p_x + \sqrt{a^2 - \frac{1}{8}}\, p_z + ad_{zx} + \sqrt{\frac{3}{8} - a^2}\, d_{x^2-y^2} + \frac{1}{\sqrt{8}}d_{z^2}$$

$$t_6 = \frac{1}{\sqrt{8}} s - \sqrt{\frac{1}{2} - a^2}\, p_x + \sqrt{a^2 - \frac{1}{8}}\, p_z - ad_{zx} + \sqrt{\frac{3}{8} - a^2}\, d_{x^2-y^2} + \frac{1}{\sqrt{8}}d_{z^2}$$

$$t_7 = \frac{1}{\sqrt{8}}\,s - \sqrt{\frac{1}{2} - a^2}\,p_y - \sqrt{a^2 - \frac{1}{8}}\,p_z + ad_{zx} - \sqrt{\frac{3}{8} - a^2}\,d_{x^2-y^2} + \frac{1}{\sqrt{8}}\,d_{z^2}$$

$$t_8 = \frac{1}{\sqrt{8}}\,s + \sqrt{\frac{1}{2} - a^2}\,p_y - \sqrt{a^2 - \frac{1}{8}}\,p_z - ad_{zx} - \sqrt{\frac{3}{8} - a^2}\,d_{x^2-y^2} + \frac{1}{\sqrt{8}}\,d_{z^2}$$

(with $1/\sqrt{8} \leqslant a < \sqrt{3/8}$) $t_1 = t_2 = t_3 = t_4 \neq t_5 = t_6 = t_7 = t_8$

If $a = 1/\sqrt{8}$, the disposition of hybrids corresponding to same nuclear charge is the following

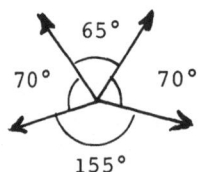

A symmetry with respect to z = 0 and a rotation of 90° about the z axis give the other hybrid orbitals.

b) sp_3d_4 (square antiprism)

$$t_1 = \frac{1}{\sqrt{8}}\,(s + p_x + p_y + p_z) + \frac{1}{2}\,d_{xy} + \frac{1}{\sqrt{8}}\,(d_{yz} + d_{zx})$$

$$t_2 = \frac{1}{\sqrt{8}}\,(s - p_x + p_y + p_z) - \frac{1}{2}\,d_{xy} + \frac{1}{\sqrt{8}}\,(d_{yz} - d_{zx})$$

$$t_3 = \frac{1}{\sqrt{8}}\,(s - p_x - p_y + p_z) + \frac{1}{2}\,d_{xy} + \frac{1}{\sqrt{8}}\,(-d_{yz} - d_{zx})$$

$$t_4 = \frac{1}{\sqrt{8}}\,(s + p_x - p_y + p_z) - \frac{1}{2}\,d_{xy} + \frac{1}{\sqrt{8}}\,(-d_{yz} + d_{zx})$$

$$t_5 = \frac{1}{\sqrt{8}}\,s + \frac{1}{2}\,p_x - \frac{1}{\sqrt{8}}\,p_z - \frac{1}{2}\,d_{zx} + \frac{1}{2}\,d_{x^2-y^2}$$

$$t_6 = \frac{1}{\sqrt{8}}\,s + \frac{1}{2}\,p_y - \frac{1}{\sqrt{8}}\,p_z - \frac{1}{2}\,d_{yz} - \frac{1}{2}\,d_{x^2-y^2}$$

$$t_7 = \frac{1}{\sqrt{8}}\,s - \frac{1}{2}\,p_x - \frac{1}{\sqrt{8}}\,p_z + \frac{1}{2}\,d_{zx} + \frac{1}{2}\,d_{x^2-y^2}$$

$$t_8 = \frac{1}{\sqrt{8}}\,s - \frac{1}{2}\,p_y - \frac{1}{\sqrt{8}}\,p_z + \frac{1}{2}\,d_{yz} - \frac{1}{2}\,d_{x^2-y^2}$$

$$t_1 = t_2 = \ldots = t_8$$

c) p_3d_5 (square antiprism)

The expression of orbitals is obtained by substituting the d_{z^2} orbital for the s orbital in the previous expressions. This hybridization is less probable because it requires greater energy than sp_3d_4.

Hybridization s×p×d×f

$\nu = 8$

a) sp_3d_3f (cube)

$$t_1 = \frac{1}{\sqrt{8}} \ (s + p_x + p_y + p_z + d_{xy} + d_{yz} + d_{zx} + f_{xyz})$$

$$t_2 = \frac{1}{\sqrt{8}} \ (s - p_x + p_y + p_z - d_{xy} + d_{yz} - d_{zx} - f_{xyz})$$

$$t_3 = \frac{1}{\sqrt{8}} \ (s - p_x - p_y + p_z + d_{xy} - d_{yz} - d_{zx} + f_{xyz})$$

$$t_4 = \frac{1}{\sqrt{8}} \ (s + p_x - p_y + p_z - d_{xy} - d_{yz} + d_{zx} - f_{xyz})$$

$$t_5 = \frac{1}{\sqrt{8}} \ (s + p_x + p_y - p_z + d_{xy} - d_{yz} - d_{zx} - f_{xyz})$$

$$t_6 = \frac{1}{\sqrt{8}} \ (s - p_x + p_y - p_z - d_{xy} - d_{yz} + d_{zx} + f_{xyz})$$

$$t_7 = \frac{1}{\sqrt{8}} \ (s - p_x - p_y - p_z + d_{xy} + d_{yz} + d_{zx} - f_{xyz})$$

$$t_8 = \frac{1}{\sqrt{8}} \ (s + p_x - p_y - p_z - d_{xy} + d_{yz} - d_{zx} + f_{xyz})$$

$$\widehat{(t_i, Oz)} \sim 55° \qquad t_1 = t_2 = \ldots = t_8$$

b) p_3d_4f (square prism)

The $\underline{s}$ orbital of the previous expressions is substituted by the d_{z^2} orbital. $\widehat{(t_i, Oz)} \sim 45°$.

$\nu = 12$

$sp_3d_5f_3$ (the orbitals are equivalent and point toward the midpoints of the edges of a cube)

$$t_1 = \frac{1}{\sqrt{12}}\, s + \frac{1}{\sqrt{8}}\, (p_x+p_z+f_z+f_x) + \frac{1}{2}\, d_{zx} + \frac{1}{2\sqrt{6}}\, d_{z^2} + \frac{1}{2\sqrt{2}}\, d_{x^2-y^2}$$

$$t_2 = \frac{1}{\sqrt{12}}\, s + \frac{1}{\sqrt{8}}\, (p_x+p_y+f_y-f_x) + \frac{1}{2}\, d_{xy} - \frac{1}{\sqrt{6}}\, d_{z^2}$$

$$t_3 = \frac{1}{\sqrt{12}}\, s + \frac{1}{\sqrt{8}}\, (p_x-p_z-f_z+f_x) - \frac{1}{2}\, d_{zx} + \frac{1}{2\sqrt{6}}\, d_{z^2} + \frac{1}{2\sqrt{2}}\, d_{x^2-y^2}$$

$$t_4 = \frac{1}{\sqrt{12}}\, s + \frac{1}{\sqrt{8}}\, (p_x-p_y-f_y-f_x) - \frac{1}{2}\, d_{xy} - \frac{1}{\sqrt{6}}\, d_{z^2}$$

$$t_5 = \frac{1}{\sqrt{12}}\, s + \frac{1}{\sqrt{8}}\, (-p_x+p_y+f_y+f_x) - \frac{1}{2}\, d_{xy} - \frac{1}{\sqrt{6}}\, d_{z^2}$$

$$t_6 = \frac{1}{\sqrt{12}}\, s + \frac{1}{\sqrt{8}}\, (-p_x+p_z+f_z-f_x) - \frac{1}{2}\, d_{zx} + \frac{1}{2\sqrt{6}}\, d_{z^2} + \frac{1}{2\sqrt{2}}\, d_{x^2-y^2}$$

$$t_7 = \frac{1}{\sqrt{12}}\, s + \frac{1}{\sqrt{8}}\, (-p_x-p_y-f_y+f_x) + \frac{1}{2}\, d_{xy} - \frac{1}{\sqrt{6}}\, d_{z^2}$$

$$t_8 = \frac{1}{\sqrt{12}}\, s + \frac{1}{\sqrt{8}}\, (-p_x-p_z-f_z-f_x) + \frac{1}{2}\, d_{yz} + \frac{1}{2\sqrt{6}}\, d_{z^2} + \frac{1}{2\sqrt{2}}\, d_{x^2-y^2}$$

$$t_9 = \frac{1}{\sqrt{12}}\, s + \frac{1}{\sqrt{8}}\, (p_y-p_z+f_z-f_y) - \frac{1}{2}\, d_{yz} + \frac{1}{2\sqrt{6}}\, d_{z^2} - \frac{1}{2\sqrt{2}}\, d_{x^2-y^2}$$

$$t_{10} = \frac{1}{\sqrt{12}}\, s + \frac{1}{\sqrt{8}}\, (p_y+p_z-f_z-f_y) + \frac{1}{2}\, d_{yz} + \frac{1}{2\sqrt{6}}\, d_{z^2} - \frac{1}{2\sqrt{2}}\, d_{x^2-y^2}$$

$$t_{11} = \frac{1}{\sqrt{12}}\, s + \frac{1}{\sqrt{8}}\, (-p_y+p_z-f_z+f_y) - \frac{1}{2}\, d_{yz} + \frac{1}{2\sqrt{6}}\, d_{z^2} - \frac{1}{2\sqrt{2}}\, d_{x^2-y^2}$$

$$t_{12} = \frac{1}{\sqrt{12}}\, s + \frac{1}{\sqrt{8}}\, (-p_y-p_z+f_z+f_y) + \frac{1}{2}\, d_{yz} + \frac{1}{2\sqrt{6}}\, d_{z^2} - \frac{1}{2\sqrt{2}}\, d_{x^2-y^2}$$

with $f_x = f_{x(y^2-z^2)} \cdots$

An example of molecular orbital localization

To illustrate molecular orbital localization in a molecular crystal, we shall consider the simple case of two neighboring H_2 molecules, whose nuclei are aligned.

Let h_i represent the <u>1s</u> atomic orbitals of H atoms. We shall use the Hückel method [7]. Let α = the monocentric integral corresponding to <u>h</u> orbitals and β, the bond integral corresponding to 1-2 and 3-4 couples, and $\lambda\beta$, the integral corresponding to 2-3 couples (see App. A).

The secular Eq. (73)

$$\begin{vmatrix} \alpha-e & \beta & 0 & 0 \\ \beta & \alpha-e & \lambda\beta & 0 \\ 0 & \lambda\beta & \alpha-e & \beta \\ 0 & 0 & \beta & \alpha-e \end{vmatrix} = 0$$

gives for the ground state, the delocalized molecular orbitals

$$\begin{cases} \varphi_1 = \dfrac{h_1 + h_4}{\left[\sqrt{4+\lambda^2}\left(\sqrt{4+\lambda^2}+\lambda\right)\right]^{1/2}} + \dfrac{1}{2}\left[\dfrac{\sqrt{4+\lambda^2}+\lambda}{\sqrt{4+\lambda^2}}\right]^{1/2} (h_2 + h_3) \\[4mm] \varphi_2 = \dfrac{h_1 - h_4}{\left[\sqrt{4+\lambda^2}\left(\sqrt{4+\lambda^2}-\lambda\right)\right]^{1/2}} + \dfrac{1}{2}\left[\dfrac{\sqrt{4+\lambda^2}-\lambda}{\sqrt{4+\lambda^2}}\right]^{1/2} (h_2 - h_3) \end{cases}$$

If the 2-3 distance is large ($\lambda \ll 1$), these orbitals become

$$\begin{cases} \varphi_1 = \dfrac{1}{2}\left[(1 - \tfrac{\lambda}{4})h_1 + (1 + \tfrac{\lambda}{4})h_2 + (1 + \tfrac{\lambda}{4})h_3 + (1 - \tfrac{\lambda}{4})h_4\right] \\[4mm] \varphi_2 = \dfrac{1}{2}\left[(1 + \tfrac{\lambda}{4})h_1 + (1 - \tfrac{\lambda}{4})h_2 - (1 - \tfrac{\lambda}{4})h_3 - (1 + \tfrac{\lambda}{4})h_4\right] \end{cases}$$

If we make the combinations $(\varphi_1 \pm \varphi_2)/\sqrt{2}$, we obtain new molecular orbitals

$$\begin{cases} \varphi_1' = \frac{1}{\sqrt{2}} (h_1 + h_2) + \frac{\lambda}{4\sqrt{2}} (h_3 - h_4) \\ \\ \varphi_2' = \frac{1}{\sqrt{2}} (h_3 + h_4) + \frac{\lambda}{4\sqrt{2}} (h_2 - h_1) \end{cases}$$

These orbitals are localized between the 1-2 and 3-4 nuclei, respectively.

For two isolated H_2 molecules, the molecular orbitals are, respectively,

$$\frac{1}{\sqrt{2}} (h_1 + h_2) \quad \text{and} \quad \frac{1}{\sqrt{2}} (h_3 + h_4) \ .$$

We see that, if the two groups of nuclei (1-2 and 3-4) are sufficiently distant, λ is weak, and the localized molecular orbitals are identical to orbitals of isolated molecules.

The electron charges ($q_r = 2 \sum_i c_{ir}^2$) corresponding to various atoms are equal to unity, whatever the value of λ.

Appendix D

Mineral hardness - Improved Mohs scale

The hardness of a mineral is an immediately accessible character, useful in its identification. One can say that A is harder than B, if A scratches B. The hardness is defined in comparison to reference minerals.

The most frequently used scale is the one proposed in 1810 by Mohs. This scale assigns integer values (from 1 to 10) to usual minerals (see Table 21). A mineral that streaks fluorite, but that is scratched by apatite, has a hardness conventionally noted as 4.5.

Other scales, based on absolute physical criteria, have been proposed: for example, the minimum pressure that should be exerted on a hand point so that its displacement scratches the surface of the sample (sclerometry), or number of turns required by a drill to obtain an impression of a given size. Such determinations can be made only in laboratory. Therefore, the Mohs scale remains the one most frequently used.

Nevertheless, this scale can be improved by means of smoothing, i.e., taking the absolute determinations into account. The hardness of reference minerals is weakly modified [52]. Table 21 gives the classical Mohs values and the smoothed values H. In this lecture, we use only the H scale.

Table 21

Mineral	Classical Mohs hardness	Improved hardness
Talc	1	1.0
Graphite	2	2.0
Calcite	3	3.2
Fluorite	4	3.7
Apatite	5	5.2
Orthoclase	6	6.0
Quartz	7	7.0
Topaz	8	8.2
Corundum	9	8.9
Diamond	10	10.0

References

[1] Pauling, L.: The Nature of Chemical Bond. Ithaca: Cornell 1940.

[2] Born, M., Oppenheimer, J.R.: Ann.d.Phys. **84**, 457 (1927).

[3] Roothaan, C.C.J.: Rev.Mod.Phys. **23**, 69 (1951).

[4] Fock, V.: Z.Physik **61**, 126 (1930).

[5] Julg, A.: Bull.Soc.Chim.Belg. **85**, 985 (1976).

[6] Rüdenberg, K.: in Modern Quantum Chemistry (Istanbul Lectures)
    ed. by Sinanoglu, O., Part I, p.85. New-York: Academic
    Press 1965.

[7] Julg, A.: Chimie Théorique. Paris: Dunod 1964.

[8] Van Vleck, J.H., Shermann, A.: Rev.Mod.Phys. **7**, 168 (1935).

[9] Julg, A., Julg, O.: Exercices de Chimie Quantique. Paris: Dunod 1967.

[10] Claverie, P., Diner, S.: in Localization and Delocalization in
     Quantum Chemistry, vol.II, p.395, ed. by Chalvet, O., et al.
     Dordrecht: Reidel Publ. 1976.

[11] Julg, A.: Chimie Quantique Structurale et éléments de Spectroscopie.
     Alger: Off.Univ.Publ. 1978.

[12] von Niessen, W.: Theoret.Chim.Acta **38**, 9 (1975).

[13] Julg, A., Marinelli, F., Pellegatti, A.: Int.J.Quant.Chem. in press.

[14] Klages, F.: Chem.Ber. **82**, 358 (1949). - Franklin, J.L.: Ind. Eng.
     Chem. **41**, 1070 (1949).

[15] Sudgen, T.M.: Nature **160**, 367 (1947).

[16] Wells, A.F.: Structural Inorganic Chemistry, 2th ed. Oxford:
     Clarendon Press 1952.

[17] Cotton, A., Wilkinson, G.: Advanced Inorganic Chemistry, 3th ed.
     New-York: J.Wiley 1972.

[18] Caillet, J., Claverie, P., Pullman, B., Acta Cryst. (in press).

[19] Julg, A.: in Topics in Current Chemistry, vol.58, p.1. Berlin:
     Springer-Verlag 1975.

[20] Hoare, M.R., Pal, P.: Adv. in Phys. **24**, 645 (1975).

[21] Crowe, R.W., Santry, D.P.: Chem.Phys. **2**, 304 (1973).

[22] Julg, A., Marinelli, F.: Int.J.Quant.Chem. **X**, 1037 (1976).

[23] Goodenough, J.B.: Les oxydes des métaux de transition (translated by Casalot, A.) Paris: Gauthier-Villars 1973.

[24] Löwdin, P.O.: J.Chem.Phys. 19, 1579 (1951).

[25] Stoll H., Preuss, H.: Int.J.Quant.Chem. IX, 775 (1975).

[26] Löwdin, P.O.: Phys.Rev. 97, 1509 (1955). - des Cloizeaux, J.: J.Phys. Radium 20, 606, 751 (1959).

[27] Di Vincenzo, T.M., Kalkan, R., Girifalco, L.A.: Phys.Rev.B 9, 3180 (1974). - Kalkan, R., Girifalco, L.A., Rothwarf, A.: Phys. Rev.B 9, 3187 (1974).

[28] Raines, S.: Proc.Phys.Soc. A 67, 52 (1954).

[29] Calais, J.L., Sperber, G.: Int.J.Quant.Chem. VII, 501 (1973).

[30] Julg, A., Del Re, G., Barone, V.: Phil.Mag. 35, 517 (1977).

[31] Mulliken, R.S.: J.Phys.Chem. 56, 295 (1952).

[32] Moore, C.: Atomic Energy Levels. Washington: Nat.Bur.Stand. Wash. 1949. - Politzer, P.: Trans. Faraday Soc. 549, 2241 (1968). - Hotop, H., Benvett, R.A.: J.Chem.Phys. 58, 2373 (1973).

[33] Verhaegen, G., Stafford, F.E., Goldfinger, P., Ackerman, M.: Trans. Faraday Soc. 58, 1926 (1962).

[34] Coulson, C.A., Redei, L.B., Stocker, D.: Proc.Roy.Soc. A 270, 357 (1962).

[35] Harrison, W.A.: Phys.Rev.B 8, 4487 (1973).

[36] Grimley, T.B.: Proc.Phys.Soc.(London) 70 A, 123 (1957).

[37] Messmer, R.P.: Chem.Phys.Letters 11, 589 (1971). - Messmer, R.P., Tucker, C.W., Johnson, K.H.: Chem.Phys.Letters 36, 423 (1975).

[38] Julg, A., Allouche, A.: Surface Science 71, 719 (1978).

[39] Yip, K.L., Fowler, W.B.: Phys.Rev.B 10, 1400 (1974). - Collins, G.A.D., Cruisksrank, D.M.J., Breeze, A.: J.Chem.Soc. Faraday Trans. II 68, 1189 (1972).

[40] Chelikowsky, J.R., Schlüter, M.: Phys.Rev.B 15, 4020 (1977).

[41] Löwdin, P.O.: Thesis, Uppsala: Almquist and Wiksells 1948. - Adv. in Phys. 5, 1 (1956). - in Aspects de la Chimie Quantique Contemporaine. Paris: Ed. C.N.R.S. (1971). - Calais, J.L., Vallin, J., Mansikka, K.M., Petterson, G.: Ark.Fys. 34, 199, 361, 371 (1967).

[42] Andzelm, J., Piela, L.: J.Phys. $\underline{C}$, Solid State Phys. $\underline{10}$, 2269 (1977).

[43] Ghio, C., Scrocco, E., Tomasi, J.: in Environmental effect on molecular structure and properties, ed. by Pullman, B., p.329. Dordrecht: Reidel Publ. 1976.

[44] Witte, H., Wölfel, E.: Z.Phys.Chemie $\underline{3}$, 296 (1955).

[45] Szigeti, B.: Trans.Faraday Soc. $\underline{45}$, 155 (1949).

[46] Hardy, J.R.: Phil.Mag. $\underline{6}$, 27 (1961). - Woods, A.D.B., Cochran, W., Brockhouse, B.N.: Phys.Rev. $\underline{119}$, 980 (1960). - Lachhman Dass, Kachkava, C.M.: Indian J.Phys. $\underline{50}$, 1043 (1976).

[47] Lundquist, S.O.: Arkiv.Fys. $\underline{9}$, 435 (1955).

[48] Jørgensen, C.K., Berthon, H., Balsenc, L.: J. of fluorine Chem. $\underline{1}$, 327 (1971).

[49] Jørgensen, C.K.: private communication.

[50] Baldock, G.K.: Proc.Phys.Soc. $\underline{A}$ $\underline{66}$, 2 (1953).

[51] Desjonquères, M.C., Cyrot-Lackmann, F.: J.Chem.Phys. $\underline{64}$, 3707 (1976).

[52] Julg, A.: Phys. and Chem. Minerals, in press.

[53] Picus, G., Burnstein, E., Henvis, B.W., Haas, M.: J.Phys.Chem.Solids $\underline{8}$, 282 (1959).

[54] Dana's Textbook of Mineralogy, ed. by Ford, W.E., New-York: J.Wiley, 4th ed 1958.

[55] Handbook of Chemistry and Physics. Cleveland: The Chemical Rubber Co, 48th ed 1967.

[56] Barinsky, R.L., Kulikova, I.M.: Phys. and Chem. Minerals $\underline{1}$, 325 (1977).

[57] Barriol, J.: Eléments de Mécanique Quantique, p.297. Paris: Masson 1966.

[58] Born, M., Mayer, J.E.: Z.f.Physik $\underline{75}$, 1 (1932).

[59] Julg, A., Létoquart, D.: Nouveau J. de Chimie $\underline{1}$, 261 (1977).

[60] Van Oosterhout, A.B.: J.Chem.Phys. $\underline{67}$, 2412 (1977).

[61] Ozias, Y., Bonnet, M.: Comp.Rend. $\underline{264}$ $\underline{C}$, 1934 (1967).

[62] Julg, A., Julg, O.: Theoret.Chim.Acta $\underline{22}$, 353 (1971).

[63] Wood, J.: Chem.Phys.Letters $\underline{50}$, 129 (1977).

[64] Goldsmith, V.M.: Ber.deut.Chem.Ges. $\underline{60}$, 1263 (1927).

[65] Ladd, M.F.C.: Theoret.Chim.Acta $\underline{12}$, 333 (1968).

[66] Wyckoff, R.W.G.: Crystal Structures. New-York: Interscience, 2th ed 1963.

[67] Ohno, K.: Theoret.Chim.Acta $\underline{2}$, 219 (1964).

[68] Pellegatti, A.: J.Chim.Phys. $\underline{68}$, 791 (1971).

[69] Julg, A.: J.Chim.Phys. $\underline{55}$, 413 (1958).

[70] Julg, A., Bénard, M., Bourg, M., Gillet, M., Gillet, E.: Phys.Rev.B $\underline{9}$, 3248 (1974).

[71] Ross, M., Mc Mahan, A.K.: Phys.Rev.B $\underline{13}$, 5154 (1976).

[72] Silver, D., Stevens, R.M.: J.Chem.Phys. $\underline{59}$, 3378 (1973).

[73] Legros, B.: Thèse, Toulouse 1972.

[74] Ducastelle, F., Cyrot-Lackmann, F.: J.Phys.Chem.Solids $\underline{32}$, 285 (1971).

[75] Braunbeck, W.: Naturw. $\underline{16}$, 546 (1928).

[76] Lennard-Jones, J.E., Dent, B.M.: Proc.Roy.Soc. London A $\underline{121}$, 247 (1948).

[77] Zwicky, F.: Phys.Z. $\underline{24}$, 131 (1923).

[78] Laramore, G.E., Switendik, A.C.: Phys.Rev.B $\underline{7}$, 3615 (1973).

[79] Duke, C.B.: Materials Sc. and Eng. $\underline{25}$, 13 (1976).

[80] Joyce, B.A.: Surface Science $\underline{35}$, 1 (1973).

[81] Schlüter, M., Chelikowsky, J.R., Lonie, S.G., Cohen, M.L.: Phys.Rev. B $\underline{12}$, 4200 (1975).

[82] Lublinsky, A.R., Duke, C.B., Lee, B.W., Mark, P.: Phys.Rev.Letters $\underline{36}$, 1058 (1976).

[83] Duke, C.B., Lublinsky, A.R., Lee, B.W., Mark, P.: J.Vac.Sc.Technol. $\underline{13}$, 189 (1976).

[84] Ricca, F., Pissani, C., Garrone, E.: J.Chem.Phys. $\underline{51}$, 4079 (1969). - Ricca, F., Garrone, E.: Faraday Soc. $\underline{66}$, 959 (1970).

[85] Randič, M., Maksič, Z.B.: Chem.Rev. $\underline{72}$, 43 (1972).

[86] Rastelli, A., Pozzoli, A.S.: J.Chem.Soc. $\underline{69}$, 256 (1973).

[87] Del Re, G.: to be published, personal communication.

[88] Allouche, A.: Thèse, Marseille 1978.

[89] Biċzo, G.: personal communication.

[90] Julg, A., Del Re, G., Barone, V.: Phil.Mag. $\underline{35}$, 517 (1977).

[91] Julg, A., Del Re, G., Bourg, M., Barone, V.: J.Phys. 38 C2, 29 (1977).

[92] Julg, A., Del Re, G., Barone, V., Montella, N.: to be published.

[93] Parr, R.G., Crawford, B.L.: J.Chem.Phys. 16, 526 (1948).

[94] Julg, A., Bénard, M.: Tetrahedron 24, 5575 (1968).

[95] Stoll, H., Preuss, H.: Phys.Stat.Sol. b 53, 519 (1973). - Leleyter, M.: Thèse, Orsay 1975.

[96] Verwey, E.J.W.: Rec.Trav.Chim. Pays-Bas 65, 521 (1946).
   - Welton-Cook, M.R., Prutton, M.: Surf. Science 64, 633 (1977).

[97] Julg, A., Bernière, P.: to be published.

[98] Tosi, M.P., Doyama, M.: Phys.Rev. 151, 642 (1966).

[99] Rao, K.J., Rao, C.N.R.: Phys.Stat.Sol. 28, 157 (1968).

[100] Guérin, P., Laforgue, A.: J.Phys. C 9, 34, 185 (1973); C 7, 37, 319 (1976).

[101] Kersten, R.: Solid State Comm. 8, 167 (1970). - Ong, C.K., Vail, J.M.: Phys.Rev. B 15, 3898 (1977).

[102] Del Re, G., Julg, A., Barone, V., Montella, N.: to be published.

[103] Gibbs, J.W.: Collected works on Thermodynamics, vol.I. New-Haven: Yale Univ. Press 1948.

[104] Wulff, G.: Z.Krist. 34, 449 (1901).

[105] Taylor, J.W.: J.Inst.Met. 83, 143 (1954).

[106] Lang, N.D., Kohn, W.: Phys.Rev.B 3, 4555 (1970).

[107] Huang, K., Wyllie, G.: Proc.Phys.Soc. A 62, 180 (1949).

[108] Stratton, R.: Phil.Mag. 44, 1247 (1953).

[109] Skapki, A.S.: Acta Metal. 4, 576 (1956).

[110] Kern, R.: Croissance des cristaux in Encyclopoedia Universalis, vol.5, p.124. Paris: Encycl. Univ. France 1968.

[111] Gordon, M.B.: Thèse 3ème cycle, Grenoble 1978 .

[112] Gillet, M.: Surface Science 67, 139 (1977).

[113] Van Arkel, A.E., de Boer, J.H.: Die chemische Bindung als elektrostatische Erscheinung. Leipzig 1931.

[114] Julg, A., Létoquart, D.: Phil.Mag. 33, 721 (1976).

[115] Pariser, R., Parr, R.G.: J.Chem.Phys. 21, 466, 767 (1953).

[116] Michaelson, H.B.: J.Applied Physics <u>48</u>, 4729 (1977); IBM Journal of Research and Development <u>22</u>, 72 (1978).

[117] Julg, A.: to be published.

Subject Index